About This Book

Why Is *Developing Technical Training* Important?

This is a book about how to develop custom workforce training programs for delivery in either classrooms or by synchronous or asynchronous e-learning. In an industry that invests over $50 billion annually in the United States alone, effective workforce training is essential! To create effective training, I recommend that you follow a systematic approach called the Instructional Systems Design (ISD) process. In this book, you will learn how to identify and teach your technical content—defined as facts, concepts, processes, procedures, and principles.

Why should you care about your content types? There are different proven instructional techniques for each of these types. By using these techniques you create learning environments that lead to the most efficient and effective acquisition of new knowledge and skills. These techniques are essentially the same whether you are developing a student manual for classroom training or screens for e-learning. In each chapter I define the content type, describe the unique teaching techniques for each type, and illustrate how to implement those techniques for either classroom or e-learning delivery. If you are new to design and development of workforce learning programs, you will be able to immediately apply these training methods to your instructional assignments.

What's New in the Third Edition?

Since its first release in the late 1980s, *Developing Technical Training* has been a popular resource for both subject-matter experts tasked with training assignments and for new training specialists. Since the second edition, written seven years ago, we have witnessed considerable evolution of digital learning environments. Synchronous e-learning was virtually unheard of at the time of the second edition. Asynchronous e-learning—known at that time as "computer-based training"—was present but accounted for a relatively small proportion of technical training. Improvements in authoring software, the rapid dominance of the Internet, the ubiquity of mobile digital devices, along with instructional cost and time savings, are among the factors promoting a growing e-learning market share. In this edition, you will find new e-learning examples that reflect how instructional methods are best implemented in virtual classroom training environments as well as in asynchronous e-learning and classroom manuals.

What Can You Achieve with This Book?

Whether you are a subject-matter expert who is assisting with a training program or a new training specialist, this book includes all the basics you need to develop a successful custom workforce training program. In particular the guidelines and examples in this book will help you:

- Follow a systematic process to define your training content and develop your training materials
- Define the four essential ingredients of your training, including the content, the learning objectives, the instructional methods, and the delivery media
- Identify the five main content types (facts, concepts, processes, procedures, and principles) that make up your technical training
- Identify the key instructional methods, including practice exercises you need to teach facts, concepts, processes, procedures, and principles
- Display important instructional methods in classroom manuals or on screens for synchronous and asynchronous e-learning
- Design powerful practice exercises that will lead to learning
- Define your content and organize your lessons
- Construct test questions to match your learning objectives and type of content

How Is This Book Organized?

This book is organized into three sections:

Section 1: An Introduction to the Technology of Training

Chapters 1 and 2 lay the foundation for the book by introducing the four ingredients of instruction, the instructional systems design process, and the five key content types that make up your training program.

Section 2: The Content-Performance Matrix

Chapters 3 through 7 make up the "heart" of this book, with each chapter focusing on one of the five main content types. Each chapter defines the content type and shows you the best instructional methods to use to teach it both for classroom and e-learning delivery.

Section 3: How to Organize Your Lessons and Exploit e-Learning Features

Chapters 8 and 9 provide more details on how to conduct a detailed job analysis and organize the content in your lessons as well as unique issues to consider when planning e-learning.

About Pfeiffer

Pfeiffer serves the professional development and hands-on resource needs of training and human resource practitioners and gives them products to do their jobs better. We deliver proven ideas and solutions from experts in HR development and HR management, and we offer effective and customizable tools to improve workplace performance. From novice to seasoned professional, Pfeiffer is the source you can trust to make yourself and your organization more successful.

Essential Knowledge Pfeiffer produces insightful, practical, and comprehensive materials on topics that matter the most to training and HR professionals. Our Essential Knowledge resources translate the expertise of seasoned professionals into practical, how-to guidance on critical workplace issues and problems. These resources are supported by case studies, worksheets, and job aids and are frequently supplemented with CD-ROMs, websites, and other means of making the content easier to read, understand, and use.

Essential Tools Pfeiffer's Essential Tools resources save time and expense by offering proven, ready-to-use materials—including exercises, activities, games, instruments, and assessments—for use during a training or team-learning event. These resources are frequently offered in looseleaf or CD-ROM format to facilitate copying and customization of the material.

Pfeiffer also recognizes the remarkable power of new technologies in expanding the reach and effectiveness of training. While e-hype has often created whizbang solutions in search of a problem, we are dedicated to bringing convenience and enhancements to proven training solutions. All our e-tools comply with rigorous functionality standards. The most appropriate technology wrapped around essential content yields the perfect solution for today's on-the-go trainers and human resource professionals.

Essential resources for training and HR professionals

To all workforce learning professionals—often under-recognized contributors to organizational performance improvement and workforce well-being.

In memory of my grandparents, whose lives modeled independence, excellence in work achievement, and family values: Dr. Oral B. Bolibaugh and Ruth Allen Bolibaugh.

Pfeiffer™

Developing
Technical
Training

A Structured Approach for Developing
Classroom and Computer-Based
Instructional Materials
Third Edition

Ruth Colvin Clark

John Wiley & Sons, Inc.

Published by Pfeiffer
An Imprint of Wiley
989 Market Street, San Francisco, CA 94103-1741
www.pfeiffer.com

Readers should be aware that Internet websites offered as citations and/or sources for further information may have changed or disappeared between the time this was written and when it is read.

For additional copies/bulk purchases of this book in the U.S. please contact 800-274-4434.

Pfeiffer books and products are available through most bookstores. To contact Pfeiffer directly call our Customer Care Department within the U.S. at 800-274-4434, outside the U.S. at 317-572-3985, fax 317-572-4002, or visit www.pfeiffer.com.

Pfeiffer also publishes its books in a variety of electronic formats. Some content that appears in print may not be available in electronic books.

Library of Congress Cataloging-in-Publication Data

Clark, Ruth Colvin.
    Developing technical training : a structured approach for developing classroom and computer-based instructional materials / Ruth Colvin Clark. —3rd ed.
        p. cm.
    Includes bibliographical references and index.
    ISBN 978-0-7879-8846-3 (cloth)
        1. Technology—Study and teaching.   2. Computer-assisted instruction.   3. Instructional systems—Design.
4. Employees—Training of.   I. Title.
    T65.C615 2007
    658.3'124—dc22

                                                                                    2007033229

Acquiring Editor: Matt Davis
Director of Development: Kathleen Dolan Davies
Production Editor: Dawn Kilgore
Editor: Rebecca Taff
Editorial Assistant: Julie Rodriquez
Manufacturing Supervisor: Becky Morgan

Printed in the United States of America

Printing   10 9 8 7 6 5 4 3 2 1

CONTENTS

ACKNOWLEDGMENTS

IN THIS NEW EDITION, I have updated and revised the examples included in the previous editions. I thank Karen Zwick of 1st Class Solutions for her assistance with classroom manual examples and Ann Kwinn of Clark Training for a revision of the Nerd example. I also appreciate the assistance or permission of the following individuals and organizations for creating or giving me access to their examples:

Alan Koenig, who programmed all of the database asynchronous e-learning examples.

Frank Nguyen, who designed and programmed the asynchronous Excel examples.

Susan Lajoie, who gave permission to use Bioworld examples.

Moody's Financial Services, who gave permission to use examples from their loan simulation software.

Finally, I am grateful to support from the Pfeiffer team, and especially to Matt Davis for editorial support.

Introduction

GETTING THE MOST FROM THIS RESOURCE

Purpose

In spite of a huge annual investment in workforce learning, all too often technical learning environments fall far short of their potential to improve organizational performance. From customer service to engine maintenance training, the guidelines in this book will help you define and develop learning events that will improve worker confidence and job performance. You will learn to define your technical content as facts, concepts, processes, procedures, or principles. Then you will see how to use proven instructional methods to teach each of these content types in the classroom, as well as via e-learning.

Audience

If you are a subject-matter expert with part-time training assignments or a new training specialist, this book is for you! I have written this book primarily for beginners who need a structured process to plan and develop customized training programs for their organizations.

Package Components

Each chapter includes a short exercise to help you apply the ideas of the chapter. The exercises are found in the appendix and are followed by solutions. In addition, I include a glossary and references to classic and more recent books and articles for those interested in more details on given topics.

An Introduction to the Technology of Training

Chapter 1: The Technology of Training

Provides an overview of the instructional systems design (ISD) process and introduces the four basic components of all training programs: the content, the objectives, the instructional methods, and the delivery media.

Chapter 2: An Introduction to Structured Lesson Design

Introduces the general structure of a lesson and the content-performance matrix. Provides the rationale for documenting the key instructional methods in the learning materials.

Figure 1.1. An Instructional Systems Design (ISD) Process

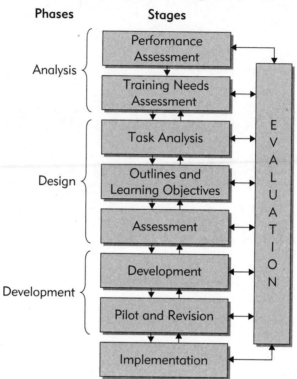

1

The Technology of Training

AN INTRODUCTION

CHAPTER OVERVIEW

An economy dependent on design, engineering, analysis, and service—in other words *on knowledge work*—cannot afford ineffective or inefficient training. Training with organizational payoff won't happen by accident. It requires a systematic approach to analyze requirements, define instructional ingredients, and create a learning environment that achieves your goals. This systematic approach is called *Instructional Systems Design* or ISD for short. The result of ISD is the definition of four main ingredients in your training program: *instructional content, learning outcomes, instructional methods, and delivery media.*

This book is about the processes and guidelines you need to develop technical training that is consistent and effective. I define technical training as learning environments delivered in face-to-face classrooms or via computer designed to build job-relevant knowledge and skills that improve bottom-line organizational performance.

The Costs of Training Waste

It's a common and costly myth that if there are ten to fifty people in a room with an "instructor" at the front showing slides and talking, learning is taking place. In other words, a training "event" is assumed to result in learning. It is further assumed that learning translates into improved job performance. Another pervasive myth suggests that training delivered on a computer is not as effective as face-to-face learning. Whether delivered in a classroom or on a computer, often training events fail to realize their potential! Participants are unable to do anything new or different after training when they return to the job. Or if they can do new and different things, those things don't translate into job skills that align to bottom-line organizational objectives. In fact, some studies have shown that learners were better off before the training than afterward, when they felt confused and inadequate about their own abilities.

Exact estimates of training waste are difficult, since training results are so rarely measured that no one really knows for sure what has—or has not—been accomplished. Only about 50 percent of companies measure learning outcomes from training, and less than a fourth make any attempt to assess job transfer or work improvement resulting from training (Sugrue & Rivera, 2005).

The costs of ineffective training are twofold. First, there are the visible dollars invested in instructors, training materials, and training administration. This is not a trivial sum. The annual *Training* magazine industry survey reports that in 2006 over $56 billion were invested by U.S. organizations in training (Industry Report, 2006). And this is a low estimate because it does not include the most expensive element of any training program—the time workers spend in training events. When training funds are not well invested, the result is waste—not only of the training expenditures, but also from lost-opportunity costs of a workforce that lacks the skills they need to fully utilize the technologies or techniques required by their jobs.

A typical lost-opportunity scenario is associated with the development and installation of a new software system. Months, even years, of effort and hundreds of thousands of dollars are invested in the design and development of the software. Then, sometimes almost as an afterthought, someone is asked to put together a training package for the end users. Because the resulting training is suboptimal, the software ends up underutilized and a portion—sometimes a substantial portion—of the system potential is never realized. Some new users ask for help from their colleagues in adjacent cubes. Others spend hours poring over confusing technical manuals. The immediate result is learning and performance that is inconsistent and inefficient. The long-term result is underutilized and mis-utilized software.

I write this book for individuals with technical training assignments who may be new training specialists or technical experts with an instructional assignment. As job performance become increasingly knowledge-based, there is a growing and appropriate trend toward using technical experts as trainers. But this brings us to another costly training myth: the misconception that all it takes for effective training is technical expertise, combined with the years most of us spent in formal educational programs. This assumption puts an unfair burden on the experts, who are not given adequate support in the preparation and delivery of their training. It is also unfair to the employees who are supposedly "trained" and later feel demoralized because they can't apply the skills needed on their jobs. Finally, poor training cheats the organization by failing to generate a return on investment. These two assumptions are illustrated in Figure 1.1.

Why We Can't Afford Ineffective Technical Training

Five major trends make the development of the human resource through effective training a greater priority than in the past:

1. New Technology: Organizations continue to be increasingly dependent on the use of new technologies, especially information technologies, as routine business tools. While many

---

**Figure 1.1. Two Fatal Assumptions About Training That Lead to
the Graveyard of Lost Business Opportunity**

Fatal Assumption # 1
A room containing 15-50
workers and an instructor
means learning is
happening

RIP
Product
Knowledge
$ 2,350,000

Fatal Assumption # 2
All it takes for effective
Instruction is technical
content

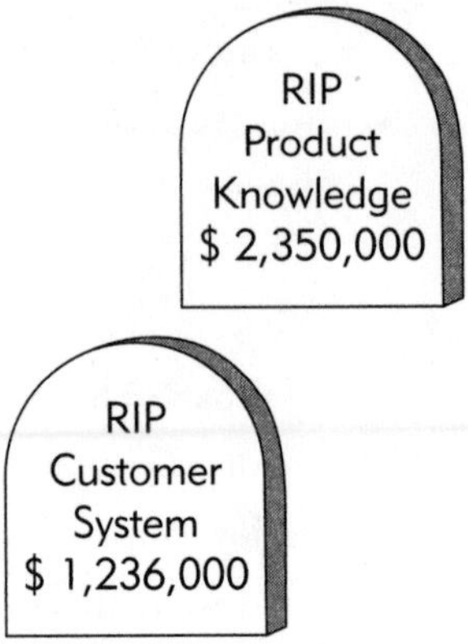

RIP
Customer
System
$ 1,236,000

tools have improved user interfaces over the past ten years,
in many cases new functionality goes unexploited in terms
of productivity payoff.

2. A Knowledge-Based Workforce: Knowledge workers have
   nearly doubled in the last half of the 20th century from
   37 percent in 1950 to nearly two-thirds of total employment in 2000 (Wolff, 2005). Reliance on a skilled workforce continues to grow in industries most dependent
   on safety, knowledge, and service. In 2005 the industry
   segments with highest per employee expenditures on training were transportation and utilities; finance, insurance,
   and real estate; and services (Sugrue & Rivera, 2005).

3. Lifelong Learning: An aging population requires organizations to think now about how to efficiently transfer a large
   skill reserve to replace a growing number of retirees. At
   the same time, new products, global competitors, updated
   policies, and emerging markets require a flexible workforce
   that can rapidly acquire and apply new skills. Lifelong
   learning requires continuous and rapid deployment of
   effective instructional resources.

4. Access to Learning Resources: The ubiquitous access to data via broadband Internet and wireless technologies makes channels of instruction broadly available to a wide population. Similarly, many workplace tools such as new software systems embed training and memory support within the tool itself. However, as we will see below, it is not the delivery medium that impacts instructional effectiveness. Only by using effective instructional methods can we harness delivery channels effectively.

5. Operational Alignment: In a global economic environment, learning must be aligned to business strategy and increasingly integrated into the work environment. Better decisions about how to deploy training resources will result in growth of "just-in-time" performance support resources and smarter use of formal training events that will be integrated as one element of larger performance improvement initiatives.

In the 21st century, the development of the human resource can no longer receive less than top priority in any organization determined to remain competitive. In fact, in a knowledge economy the emphasis shifts from traditional capital resources to the human resource for competitive edge.

If you are a technical expert, you are already a valuable resource for your skills and knowledge. But learn to transmit your expertise to others effectively and efficiently and you quadruple your value. If you are a new training specialist you will add value by learning to elicit knowledge and skills from experts and to organize and display that content in ways that lead to efficient learning and performance by the workforce. Follow the guidelines in this book and your training will enable the workforce to fully utilize the skills you teach and to feel more confident about their work tasks. Furthermore, if you follow my guidelines for measuring training outcomes, you will know—not just guess at—your training results.

What Is Technical Training?

Some interpret the term technical training as meaning "hard skills," such as using a new computer system or applying safety standards during equipment operations. I define technical training as *"a structured learning environment engineered to improve workplace performance in ways that are aligned with bottom-line business goals."* This definition includes five main elements:

1. Structured—An effective training environment is designed to optimize learning both during the training event and afterward on the job. Following a structured process and producing a structured product minimizes inconsistency in learning environments and aligns instructional products to job-essential knowledge and skills.

2. Environment—Workforce learning is moving from a series of isolated training events to environments that incorporate diverse knowledge resources such as repositories of examples, performance templates, and access to expertise, along with traditional events in face-to-face and virtual media.

3. Engineered—Effective learning environments do not happen by accident or by seat-of-the pants efforts. Effective learning environments are the products of a structured process and proven instructional methods matched to your content.

4. Workplace Performance—An effective training program starts and ends with the job. It includes guidelines, examples, and exercises that are job relevant.

5. Business Goals—An effective training program focuses on knowledge and skills that are aligned to important organizational objectives. Rather than training "communication skills," an effective learning design defines specific behaviors associated with the types of communications needed to support organizational objectives.

Technical training includes both hard and soft skills. It incorporates on-the-job performance support as well as training events—delivered in classrooms and on computers. Note that I emphasize training as a process rather than an event. Too often training is conceived and implemented as a discrete event with a beginning and end. Instead of "classes," consider learning as an ongoing process that can be engineered to include both traditional leader-led instruction in face-to-face or virtual classrooms as well as asynchronous activities and resources scheduled before, between, and after more traditional events.

Some important skill requirements such as management skills or skills associated with widespread computer programs can be achieved with "off-the-shelf" prepackaged training materials. However, many job tasks are unique to a given industry, organization, or department. No off-the-shelf training exists to meet these needs. It is this training that will be developed by or under the supervision of each organization's training staff. Industry-specific training runs the gamut from specialized computer systems to customer communication skills. It includes what we traditionally have called "hard" skills as well as soft skills. For example, in the banking or telecommunications industries the call center representative must be able to apply industry regulations, company-specific processes and policies, specialized computer applications, and communication skills to respond accurately and effectively to customer requests.

The Technology of Training

In the last part of the 20th century, training evolved from a craft to a technology. A technology is the application of scientific principles to achieve a practical and predictable result. That means the guidelines provided here go beyond a collection of experiences. Instead, they are based on research from learning psychology. The principles guiding the design of instructional materials constitute a relatively new field called *Instructional Technology*. Instructional technology takes a systematic approach to planning, developing, and evaluating training. It also offers a set of guidelines that will help you package your technical knowledge in a form that makes it learnable.

We begin with an overview of a systematic approach to defining, designing, and developing training.

Instructional Systems Design: An Overview

Out of large-scale engineering projects of the mid-20th century, systematic processes for planning, designing, and building products were born. In the production of complex products such as space stations and aircraft, it was discovered that front-end planning and design saved back-end grief in the actual production effort. More recently, data-processing specialists adopted a similar model called the Systems Life Cycle for the design of complex software. By spending up-front time in analysis and design, they avoid costly mistakes and subsequent rework during the actual construction of the software. Likewise, in the design of instructional products, a systematic process has proved much more effective than jumping right into producing learning materials. Used extensively by the military for training, *instructional systems design* (ISD) methodologies have been widely adopted by most workforce learning departments. Several texts on ISD are referenced in the For More Information section at the end of this chapter. For our purposes, a quick overview of the model will serve. If you are familiar with the ISD model, continue on to the section on the Four Ingredients of Instruction, next in this chapter.

Figure 1.2 illustrates a typical ISD model. Note the five major phases: *analysis, design, development, evaluation,* and *implementation.* The analysis and design phases, which can consume up to 50 percent of the total project effort, include stages of *performance and training needs assessment, task analysis, outlines and learning objectives,* and *assessment.* The next section will summarize each stage in the ISD process.

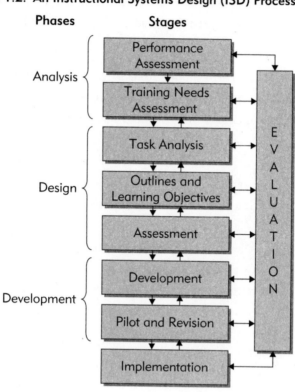

Figure 1.2. An Instructional Systems Design (ISD) Process

Performance Assessment

Often training professionals are asked to develop some type of training program when, in fact, training won't help. For example, in several recent government agency ethics scandals, the solution was ethics training. However, in and of itself, ethics training is unlikely to reduce unethical behaviors. Many factors influence workplace performance. Workers need clear job standards, feedback on results, incentives aligned to desired results, consequences for undesirable actions, usable tools and materials, and effective business processes, to name just a few. Training provides only one element of effective performance—knowledge and skills. Before jumping into a costly training design and development effort, conduct a performance analysis to ensure that, in fact, the business goal can be realized with a training effort and also to identify what in addition to training is needed. For example, in cases of ethics problems, clear performance standards, feedback, and consequences for unethical behavior are needed.

Using Performance Assessment in Customer Service.   Suppose you faced the following real-life problem. Customers were complaining that telephone service agents were rude. You are asked to develop a customer-courtesy training program. However, during the performance assessment, you note that agents are monitored and rewarded based on the number of customer calls they process during a day. Agents are required to handle at least 120 telephone calls per day and are "counseled" if performance falls below that level. The feedback-and-reward system encourages abrupt conversations in order to meet the target call volume. No amount of training will change that behavior as long as rewards are based solely on call volume. Therefore, the performance assessment report recommends several changes in the performance environment, as well as training.

Training Needs Assessment

Once you have determined that, in fact, a void of knowledge and skills does contribute to the problem or is required to meet the operational goal, you next define the specific knowledge and skills required. A training needs assessment is used to hone in on *what* training is needed by *whom* and how it should best be *delivered*.

Using Needs Assessment to Focus Credit Analysis Training.   In a large utility, credit specialists complained that customer service representatives were mishandling the credit aspects of their job. Customer service representatives did not agree with the credit department's conclusions. To define what, if any, training was needed by whom, a training needs assessment was conducted. All the credit-related documentation generated by customer service representatives for a one-week period was collected and evaluated for errors. During the same week, questions customer service representatives asked credit specialists were recorded and analyzed. As a third check, a credit skills test was administered to a sample of both customer service representatives and credit specialists. The actual credit documents, the questions asked by customer service representatives, and the items missed on the tests were analyzed for evidence of common misconceptions.

The test results showed that about 40 percent of both customer service representatives *and* credit specialists needed training on commercial accounts. The documentation evaluations and on-the-job questions confirmed the test results. The training needs assessment saved money by narrowing the scope of what needed to be trained as well as defining who actually needed training. In addition, workforce interviews determined that all employees had access to the corporate intranet and that a combination of training and working aids delivered electronically would be the most efficient approach.

In summary, the performance analysis and training needs assessment stages answer the question, "What are the best paths to improve business goals and, if training is one solution, who needs training on what, delivered how?" For more information on performance and needs assessments see the sources at the end of the chapter.

Task Analysis

Once you have determined what training is needed, your next step is to systematically define the content and outcomes of the training program. Now you will observe skilled performers, interview performers and managers, and review documentation relevant to the job. If you are a job expert already, you have much of the knowledge in your head. The problem is that much of your knowledge may be "tacit" knowledge. You have used your knowledge and skills for so long that they have become automatic. You will need to invest effort to adopt the mindset of the novice to be sure that you have included everything the unskilled worker needs. If you are a new training specialist, you will need to interview and observe experts and read related documentation as the basis for your task analysis. How to do this will be described throughout the remaining chapters.

During task analysis you not only identify the knowledge and skill requirements of the job, but you also develop a learner profile that defines the target audience's prior knowledge and skills. Research tells us that prior knowledge is the single most important individual difference that will influence learning (Clark, Nguyen, & Sweller, 2006). By subtracting the existing knowledge and skills from those required by the job, you define the content of the training program. By adjusting instructional methods for learners of higher and lower prior knowledge, you maximize instructional efficiency. More detail on how to do a task analysis is included in Chapter 8.

Outlines and Learning Objectives

While you are defining the content of the training program, you also plan the sequence of content and the required instructional methods such as practice. You also specify what the learners will do with that content by writing clearly stated learning objectives. For example, if the content of your training program was how to change a flat tire, your lesson outline would include an introduction, a section on the major tools needed to change a tire with a brief practice exercise, a section on how to change the tire with hands-on practice, and a summary. Your learning objective would require the trainees to change a tire by the end of the program. The section on the Four Ingredients of Instruction that follows will describe learning objectives in greater detail.

Assessment

You will need some way to determine that your instruction has been successful. You do this by evaluating how well the trainees have achieved the learning objectives. If your objective stated that trainees would be able to change a flat tire, you would administer an assessment to see that each trainee in fact could change a flat tire at the end of the course. To see whether your objectives have been reached, you design tests matched to the objectives. If trainees do well on the tests, you know your instruction has been successful.

Remember, I started this chapter by mentioning that many training programs cannot state what results have been achieved. That is because they often have no measurable instructional objectives. Or if objectives are available, there is no systematic attempt to see whether they have been attained. Even if the test results are not given back to the trainees, they can offer you evidence of the success of your instructional environment.

Each chapter in this book will describe appropriate testing techniques for evaluating your training effectiveness. However, the construction of useful tests is a major project in itself and requires specialized knowledge and skills. In addition, if test scores are reported to others, your organization will have legal requirements to validate the tests. If you are lucky, you may have testing experts in your organization. Check with your human resources department. Resources on testing are listed at the end of the chapter.

Development

Once you have completed task analysis and written outlines, performance objectives, and tests, you are ready to develop the instructional materials. The development phase involves the preparation of any instructional resources to be used by the learners during the training. This includes writing student workbooks, generating PowerPoint slides, developing practice exercises and case studies, or preparing video- or e-learning storyboards. This book offers guidelines to help you develop effective instructional materials for both print and computer media.

Piloting and Revision

Once you have developed a draft of your instructional materials, you will need to try them out. You will always find problems with them. Your directions may be confusing or you may not have enough practice on a particular section. Only by pilot-testing your materials will you discover the problems. You will identify problems by interviewing the pilot group of students *and* by evaluating the assessments you give them during the pilot session. Based on what your pilot students tell you and on how they do on the assessments, you will need to revise the materials to resolve the major problems encountered. Your training program will never be perfect; focus on correcting major problems that surface in your pilot quickly and efficiently. Design of effective instructional environments is always a tradeoff between quality and time efficiency.

Implementation

After the instructional program is revised, it can be implemented on a major scale. This might mean distributing your course over the intranet to 50,000 learners worldwide or teaching it in a classroom to small groups of ten or twelve. Implementation requires effective processes to market the training, to assign the right training to the right workers, and to monitor and record training completion. Large organizations rely on learning management systems to administer the implementation of training.

Evaluation

You may notice that in Figure 1.2 I align evaluation to each stage of the ISD process. That is because there are several different phases to your evaluation efforts. A convenient way to summarize evaluation is with the four levels of evaluation developed by Kirkpatrick (1994). Level 1 measures learner reactions to the training. Most training organizations capture Level 1 data through "smile sheets" distributed at the end of the event. I already discussed testing to measure learning outcomes. Learning outcomes are the focus of evaluation Level 2. Even a new skill learned in class will not necessarily be applied back in the workplace. Level 3 measures transfer of learning—the application of new knowledge and skills after the formal training event. Finally, Level 4 gets back to the bottom line. Does the training initiative result in improved organizational results and generate a return on investment?

Keep in mind that entire books have been written on each of the ISD stages I've briefly introduced above. I list several of these at the end of the chapter. My goal here is to give you an introduction to the ISD process.

Is ISD Dead?

Several articles written around the turn of the 21st century proclaimed the death of ISD due largely to the length of time required to work through the process (Gordon & Zemke, 2000; Zemke & Rossett, 2002). Looking around, however, at contemporary practices in learning organizations, it is safe to say that the death certificate was a bit premature.

What is dead is a lengthy linear approach to ISD. We have learned to be nimble by working in a circular rather than in a linear fashion. Thus rapid prototyping is used to conduct a fast job analysis and develop a first iteration "straw man" training. The straw man is then refined and elaborated with additional iterations through the ISD process. The number of iterations depends on the criticality of the training outcomes and the constraints of the situation, including time, resources, and political agendas.

With this overview of the ISD process completed, let's examine the four major ingredients of all training programs and relate them to the ISD model.

The Four Ingredients of Instruction

All training programs incorporate four major ingredients: *content, learning outcomes, instructional methods,* and *delivery media.* An effective training program carefully accounts for and deploys each ingredient to optimize results. To illustrate each ingredient, I use examples from an imaginary course on oral hygiene. The audience for the course is a group of friendly humanoid aliens who are adapting to Earth culture. These aliens are familiar with mouths and teeth but that's about all they know.

Ingredient 1: The Content

It's obvious that all training includes content. The content, or course information, is defined and organized during the design stages of the ISD process—specifically during the job analysis and the outline stages. By subtracting the knowledge and skills of the intended audience from those of the job, you can derive final course content.

This sounds easier than it often is. That's why the analysis and design of your courses can consume up to 50 percent of your total development effort. If you are a technical expert, you may be one of the major resources of knowledge and skills. As mentioned above, your major challenge will be to make all that knowledge explicit and to organize it logically. If you are a training specialist, you will need to work with experts to identify the relevant content.

All training content can be classified as one of five types: *facts, concepts, processes, procedures*, and *principles*. This book is organized around each of these types of content. Thus, Chapter 3 deals with how to teach procedures, Chapter 4 with how to teach concepts, and so forth.

As shown in Figure 1.3, our oral hygiene course includes content related to toothbrushes, knowing how often to brush, and being able to brush correctly. Each of these is a different type of content. "Toothbrush" is an example of a concept, while "how to brush" is a procedure.

While the content is important, many training projects never go beyond it in their course development. Courses end up as massive dumps of information. To avoid this pitfall, ask yourself, "What do I want my learners to do with the content?" The purpose of business training is to give employees capabilities they need to perform their jobs effectively. Therefore, defining what workers must *do* with the content is as important as defining the content itself. That brings us to the second ingredient of instruction: the learning outcomes.

Figure 1.3. Ingredient 1: Content

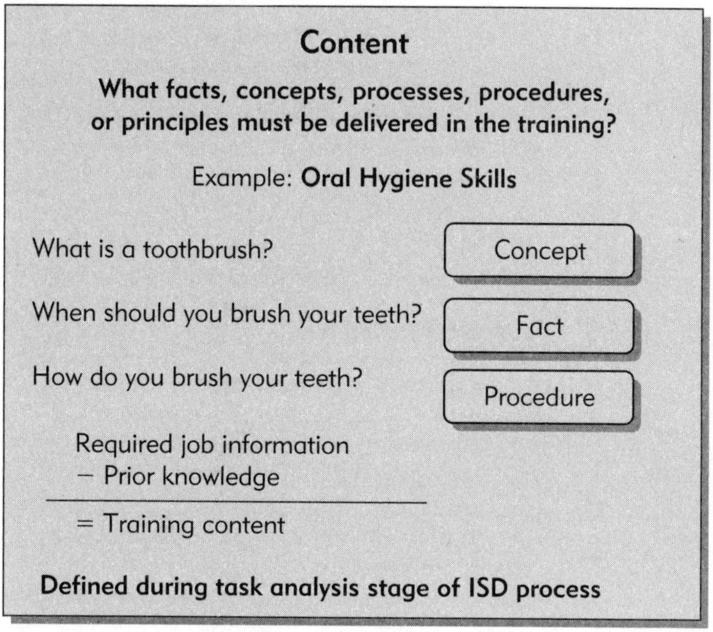

Content

What facts, concepts, processes, procedures, or principles must be delivered in the training?

Example: **Oral Hygiene Skills**

What is a toothbrush?   [Concept]

When should you brush your teeth?   [Fact]

How do you brush your teeth?   [Procedure]

Required job information
— Prior knowledge
_____

= Training content

Defined during task analysis stage of ISD process

Ingredient 2: The Learning Outcomes

The learning outcomes are clear statements of what the learners will be doing when they have achieved course or lesson goals. We call learning outcome statements *learning objectives.* Learning objectives should mirror what must be done on the job. Each of your lessons will have at least one major learning objective, and many will include supporting objectives as well.

Note the learning objectives for our oral hygiene lesson in Figure 1.4. These sample objectives include a clear action statement, a description of conditions under which the action will take place, and a standard of quality required. The first objective is a supporting or enabling objective, which describes how the learner will demonstrate that he can identify the concepts "toothbrush" and "toothpaste." The second objective is called the major or terminal lesson objective. It describes what the student will do when he has learned the procedure of brushing his teeth.

Figure 1.4. Ingredient 2: Learning Outcomes

Learning Outcomes

What must the trainees be able <u>to do</u> at the end of the training?

- Stated in learning objectives
- Mirrors what will be done on the job

Example: **Oral Hygiene Skills**

Supporting Objective:

Given bathroom supplies, learners will identify toothbrush and toothpaste with no errors.

Terminal Objective:

Given toothbrush and toothpaste, learners will clean teeth so there are fewer than three dots on the red dye test.

Defined during task analysis stage of ISD Process

Notice that these learning outcomes include an *observable* action verb. They avoid use of words such as "know" or "understand." Why? The learning outcome will be used to measure the effectiveness of the training. Suppose your outcome was "the students will *know* what a toothbrush is." How will you or the students determine that they "know"? An observable action, something we can see right away, is required. So we ask the student to pick out the toothbrush from an assortment of common bathroom supplies.

The learning objective is important because it gives you a framework for designing practice exercises and evaluating lesson success. The learning objectives, practice exercises, and test items are like jigsaw puzzle pieces. Each matches the others to make the training internally consistent. And all of them match the job to make the training valid. For example, if the objective requires the learner to brush his teeth, a practice exercise will ask participants to brush their teeth during the class and a test will observe and evaluate participants brushing their teeth.

Ingredient 3: The Instructional Methods

This book is primarily about instructional methods. Once you have identified both the content and the learning objectives, you are ready to start developing the learning materials. When developing instructional materials, use a proven set of tools known as *instructional methods*. These methods are the psychologically active ingredients of your training program that will best promote learning.

As you can see in Figure 1.5, instructional methods are of two major types: informational displays and practice exercises with feedback. The type of informational displays you will need depends on the type of content. Displays for facts are different from those needed for concepts or procedures. Each chapter in this book will describe

Figure 1.5. Ingredient 3: Instructional Methods

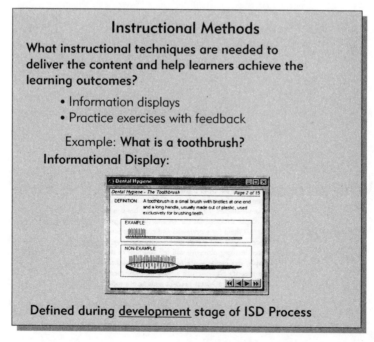

the informational displays needed for the various types of content and show you how to format them for workbooks and for e-learning.

The practice exercise will match the learning objective. A well-written learning objective tells you how to design the practice. For example, the performance outcome that states that the learner will be able to identify a toothbrush among an assortment of bathroom tools tells you to give the students practice picking out toothbrushes from a variety of common bathroom supplies. I will describe how to design effective practice exercises in each chapter, with examples drawn from a number of successful training courses.

As mentioned earlier, this book will illustrate the use of these instructional methods for two main delivery media: (1) print in the form of handouts or workbooks and (2) e-learning, both synchronous and asynchronous. This brings us to the last ingredient of instruction: the delivery media.

Ingredient 4: The Delivery Media

The instructional methods must be delivered through a medium or a blend of media. In a typical classroom, the media mix includes the instructor, a workbook or handouts, PowerPoint slides, wall charts, and perhaps a video. When designing e-learning, the computer is used to deliver visuals, animations, audio, or simulations. Many e-learning courses are supplemented by print materials. The delivery media used for our oral hygiene course includes an instructor, a workbook, and a video. (See Figure 1.6.)

Which Instructional Media Are the Best?

B.F. Skinner (1961) once wrote about teaching machines: "Obviously the machine itself does not teach. It simply brings the student into contact with the person who composed the material it presents. It is a labor-saving device because it can bring one programmer into contact with an indefinite number of students." Skinner would have been delighted to see his early teaching machine transformed into

Figure 1.6. Ingredient 4: Delivery Media

Delivery Media

What mix of media will most efficiently and effectively deliver the instructional methods?

Example: **Oral Hygiene Skills**

Delivery	Instructor - Lab
Media	Workbook
	Video

Base media decisions on:
- Which medium can deliver the required instructional methods?
- Cost effectiveness: development and delivery
- Resources: location of learners, available delivery platforms, budget, preferences

Defined during <u>needs assessment</u> stage of ISD process

today's e-learning opportunities! In spite of Skinner's early insightful observation, research studies have tried for years to identify the best media. But there are no "best" media! As Skinner stated, the media are passive carriers of the active ingredients of learning—the instructional methods. No one medium is better than another as long as it can carry the methods needed to achieve the learning objective. Thus comparisons of courses taught by an instructor with the same courses taught by computer show no differences in learning, provided the same instructional methods are used (Clark, 1994).

Media Blends

Although it is instructional methods that determine learning—not delivery media—not all media are equivalent. Not all media can carry all instructional methods. For example, a book cannot deliver audio or simulations. Select the most cost-effective mix of media that will carry the instructional methods you need to achieve your goals.

For example, effective sales training requires examples of successful exchanges between account representatives and customers. An asynchronous e-learning lesson can be used to explain and demonstrate effective sales techniques. However, role-play practice may be best implemented in an instructor-led event. Also use media blends to move from a "training event" mentality to the concept of learning as a process that extends over time and space. For example, a formal training event is supplemented by intranet resources such as sample project proposals, video examples, blogs, wikis, and coaches.

Synchronous vs. Asynchronous e-Learning

As you can see in Figure 1.7, there has been a steady decline in classroom training in workforce learning over the past six years. It looks as if we are moving toward a 50–50 mix of classroom and digital delivery. e-Learning assumes two main formats: synchronous and asynchronous. Asynchronous lessons are generally self-study, self-paced lessons designed for solo learning. In contrast, *synchronous e-learning,* also known as the *virtual classroom,* is an instructor-led event attended by learners at the same time but in different places. Examples from an Excel course presented via asynchronous and virtual classroom e-learning are shown in Figures 1.8 and 1.9.

Figure 1.7. Percentage Training Hours Delivered by Classroom and Technology

Based on data from Sugrue and Rivera, 2005

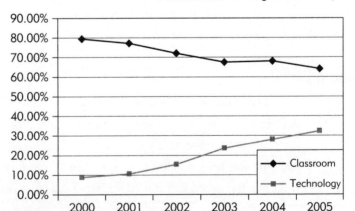

Figure 1.8. An Asynchronous e-Learning Lesson on Excel

From Clark, Nguyen, and Sweller, 2006

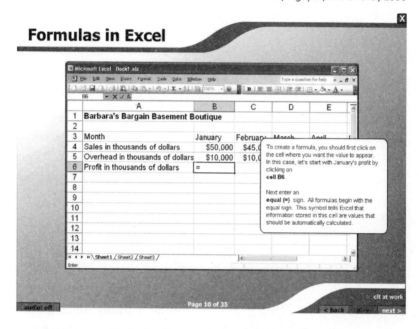

Figure 1.9. A Virtual Classroom e-Lesson on Excel

From Clark and Kwinn, 2007

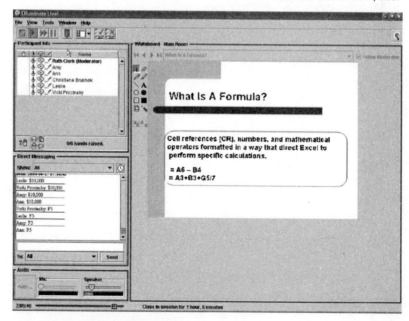

I will illustrate how to implement instructional methods in both these forms of e-learning throughout the book.

Defining Your Media Blend

During the training needs assessment, you will typically determine what blend of delivery media will best meet your specific situation. If you needed to teach a course on statistical quality control, you could choose a classroom or an e-learning delivery. If the training methods were consistently designed, learning outcomes would be equivalent from either medium and your preference would depend on an analysis of cost benefit. If the course is to be delivered to 50,000 employees who work with computers already and are located internationally, it would probably be more cost-effective to use computer-based delivery supplemented with print than to pay the expenses of sending instructors and/or students to face-to-face courses.

Many factors can influence your selection of media. Some factors are pragmatic such as availability of computers to learners, location and number of learners, time (to produce and to consume learning), budget, criticality of skills mastery, and the political landscape of your setting. Other factors are instructional such as which modes (for example, visuals—still or moving, text, or audio) are best to promote learning, or which instructional methods, such as simulations or hands-on practice, are needed. There is no one set of guidelines for your media choices. Rather, you will need to consider each situation uniquely and select a mix of media—each meeting the pragmatic and instructional constraints you identify during your training needs assessment.

Time for ISD

Time and time again I hear the comment that training professionals are asked to have an extensive training program ready to go in two or four weeks. No time is allowed for analysis or design. These time pressures are due in part to the short timelines inherent in competitive business climates. They also reflect a lack of understanding of the ISD process—its resource requirements and its benefits.

The solution to short timelines is multi-pronged. First, establish good relationships with your line clients—relationships that build trust. As part of that relationship, educate your clients about the resources and benefits from applying an ISD process. Second, be responsive. Maybe you can't provide a full-fledged learning environment by the imposed deadline. But perhaps you could provide a "first phase." The first phase might be a series of working aids to get workers started. Third, evaluate outcomes. If you can demonstrate that your rushed training efforts did not lead to desired skill levels or to confident workers, you can perhaps leverage more resources—if not on this project, on the next one. Fourth, judge when to hold and when to fold. Leverage your resources on projects that are most likely to have strategic payoff to the organization.

Check Your Understanding

To see whether you can distinguish the four ingredients of instruction, try the practice exercise for Chapter 1 in the Appendix.

COMING NEXT

An Introduction to Structured Lesson Design

Now that we have overviewed the ISD process and the four ingredients of instruction, we will zoom into more detail at the lesson level. Chapter 2 will introduce you to structured lesson design with an overview of a typical technical lesson, a summary of the value of a structured writing approach for instructional materials, and an overview of the learning taxonomy that is the foundation for Chapters 3 through 7.

For More Information

Clark, R.C., & Kwinn, A. (2007). *The new virtual classroom.* San Francisco, CA: Pfeiffer.

Clark, R.E. (1994). Media will never influence learning. *Educational Technology Research and Development, 42* (2), 21–30.

Mager, R.F. (1997). *Preparing instructional objectives* (3rd ed.). Atlanta, GA: Center for Effective Performance.

Mager, R.F. (1997). *Measuring instructional results* (3rd ed.). Atlanta, GA: Center for Effective Performance.

Mager, R.F., & Pipe, P. (1997). *Analyzing performance problems* (3rd ed.). Atlanta, GA: Center for Effective Performance.

Robinson, D.G., & Robinson, J.C. (1995). *Performance consulting.* San Francisco, CA: Berrett-Koehler.

Rothwell, W.J. (2004). *Mastering the instructional design process.* San Francisco, CA: Pfeiffer.

Rothwell, W.J. (2005). *Beyond training and development* (2nd ed.). New York: American Management Association.

Rossett, A. (1999). *First things fast: A handbook for performance analysis.* San Francisco: CA: Pfeiffer.

Shrock, S., & Coscarelli, W. (2000). *Criterion-referenced test development* (2nd ed.). Silver Spring, MD: International Society for Performance Improvement.

Zemke, R., & Kramlinger, T. (1982). *Figuring things out: A trainer's guide to needs and task analysis.* Reading, MA: Addison-Wesley.

Wolff, E.N. (2005). The growth of information workers in the U.S. economy. *Communications of the ACM, 48*(10), 37–42.

Zemke, R., & Rossett, A. (2002). A hard look at ISD. *Training, 39*(2), 26–35.

Figure 2.1. A Structured Lesson

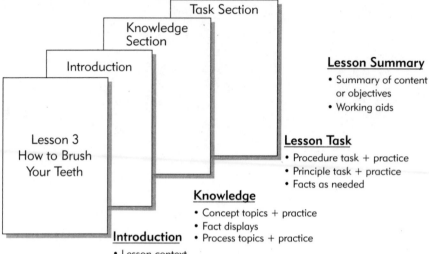

Lesson Summary
- Summary of content or objectives
- Working aids

Lesson Task
- Procedure task + practice
- Principle task + practice
- Facts as needed

Knowledge
- Concept topics + practice
- Fact displays
- Process topics + practice

Introduction
- Lesson context
- Lesson importance
- Content overview
- Lesson objectives
- Table of contents

2

An Introduction to Structured Lesson Design

CHAPTER OVERVIEW

In Chapter 1, I summarized the instructional systems design (ISD) model and described the four ingredients of all training programs: information (or content), learning outcomes, instructional methods, and the media used to deliver the methods. This book is about ingredient number three, the instructional methods. Instructional methods are the psychological tools you can use to generate the learning outcomes that best improve job performance. Ways to format these methods for two media—workbooks for instructor-led settings and computers for both synchronous and asynchronous e-learning are described in detail in Section Two, Chapters 3 through 7.

This chapter sets the context for Section Two by introducing the following principles for structured lesson design:

• A generic structure for any technical lesson

• Guidelines for three communication modes of text, graphics, and audio

• Communication guidelines for instructor-led training

• Communication guidelines for e-learning

• An Introduction to the Content-Performance Matrix, which forms the basis for the instructional methods to be presented in Section Two.

The Anatomy of a Lesson

Figure 2.1 illustrates a high-level structure that you can use for the organization of content in your lessons. It provides a consistent framework into which you can incorporate the instructional methods needed for learning. Using a structured approach has several advantages. If more than one instructor is developing a course, it will look like one course rather than a mixture of several different courses. If a series of lessons is built by different authors or even the same author over time, consistency can be maintained. Once you are familiar with the structured lesson design formats, you will save a lot of development time by plugging in the formats that fit the overall structure of your lessons. If your organization includes multiple courses that share some content, you can reuse structured materials by swapping topics among lessons written for different courses. Because the structure incorporates methods based on evidence, you can be assured of providing effective instructional materials.

Figure 2.1. The Anatomy of a Lesson

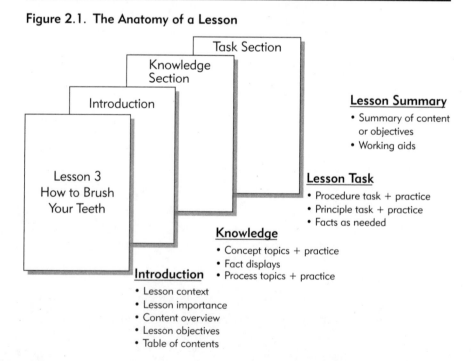

Lessons using this structured approach have three major sections:

1. Introduction,

2. Supporting knowledge, and

3. Major task(s) steps or guidelines of the lesson, including a brief lesson summary.

The supporting knowledge (Section 2) and task steps or guidelines (Section 3) make up the bulk of the lesson. Section 3 teaches the major skill or task of that lesson. Section 2 teaches all the major knowledge content needed to perform the steps or guidelines of that major skill or task. Both sections include blocks of information interspersed with practice exercises. Chapters 3 through 7 will illustrate these information displays and practice exercises. For now, let's take a brief look at each of the four sections of a technical lesson.

The Lesson Introduction

A lesson introduction typically uses one to two pages in a workbook or two to three screens in an e-learning lesson. An effective introduction both orients and motivates learners. You can refer to the "How to Brush Your Teeth" lesson introduction illustrated in Figure 2.2 as I summarize the major elements.

The introduction to the lesson should include:

1. *Context:* the first section, labeled "Introduction" in my example, orients the learner by summarizing how this lesson fits into the overall course. In my example I use a sentence. However, you might want to include a visual illustration such as a course map that shows how each lesson fits into the total structure.

2. *Importance:* an illustration of the benefits and/or importance of the lesson in order to motivate the learner. Although you can use text to state lesson importance, you can also creatively use visuals, relevant data, an in-class activity, or a discussion that makes the benefits salient.

Figure 2.2. Introduction Page from Lesson on How to Brush Teeth

How to Brush Your Teeth

Introduction	In a prior lesson you have learned about human body and hair hygiene practices. Now we will look at special care given to the teeth and mouth.
Importance	The proper care of teeth is important not only for social acceptability, but for health reasons as well. You will need to follow the guidelines in this lesson to fit into the culture.
Lesson overview	This lesson will teach you the proper tools, rules, and the procedure for brushing your teeth.
Objectives	You will • brush your teeth following the steps until they show three or less red spots from the red dye test, *and* • select a toothbrush and toothpaste from ten objects found in a bathroom.
Order of topics	This lesson contains the following topics.

Topic	SeePage
Tools for Brushing Your Teeth	51
What Is a Toothbrush?	52
What Is Toothpaste?	53
Practice1: What Is a Toothbrush and Toothpaste?	54
When to Brush Your Teeth and Practice 2	55
How to Brush Your Teeth and Practice 3	56

You may have heard the phrase: WIFM (What's in it for me?). This is where the WIFM should be placed.

3. *Content Overview:* a summary of the main content topics of the lesson. This brief overview is intended to "tell 'em what you're gonna tell 'em." It should be quite brief— enough for a preview but not so much as to bog learners down. Some instructional specialists refer to the content overview as an "advance organizer."

4. *Lesson Objectives:* an informal statement of what the learner will be able to do at the end of the lesson.

5. *Table of Contents or Navigational List of Topics.* The goal is to provide learners with quick access to topics within the lesson.

In asynchronous e-learning, your introduction should include an estimate of lesson completion time (because the computer gives no physical clues as to lesson length) and, if this is your first lesson of the course, include an orientation to features in your interface such as how to advance the screens or how to access audio or text.

The Lesson Body

Following the introduction, the body of the lesson includes two sections: *knowledge* and the *major task steps or guidelines.* In the supporting knowledge section, the learner should find all the major concepts related to the major lesson task. For example, the target task of our tooth brushing lesson is "how to brush your teeth." But, in order to brush, you would need to know about toothbrushes and toothpaste. You will use information displays and sometimes include practice exercises in this section. Technical experts often leave out critical knowledge topics because they forget that novices lack this information. Instead, they jump right into the major task, and trainees become confused by the use of terminology that was never explained. Research has proven that, rather than lumping all the knowledge together with the steps or guidelines, learning is more efficient when the major concepts are taught first, followed by the steps or guidelines that use those concepts. That way the learner is not bombarded with all of the information at one time.

The third section of your lesson teaches the steps or guidelines associated with the major task or skill to be trained. This section includes both information displays and practice exercises. The major task will usually be based on either a procedure that is described in the next chapter or on guidelines derived from principles described in Chapter 7.

The Lesson Summary

On one of your final lesson pages include a brief lesson summary, which includes a condensed presentation of the key information points and a reminder of the lesson objective to focus trainee attention on their achievement of the performance outcomes of the lesson. It might also include a preview of the next lesson.

Lesson Structure for Workbooks Versus e-Learning

The structure just described can be used for e-learning as well as for workbooks. The major differences will be the display and the spacing of information with practice within the sections. Both synchronous and asynchronous forms of e-learning must combat learner dropout—both mental and physical. To do so they must incorporate proven engagement strategies to maintain learner interest and mediate learning. The main path to engagement in e-learning is visuals as well as presentation of relatively small amounts of information, coupled with frequent response opportunities. While both classroom and e-learning must be highly interactive, in a face-to-face setting, the instructor can maintain attention for relatively longer periods of time than in computer-delivered instruction. Therefore, as we discuss design of e-learning, you will see greater use of graphics as well as small chunks of information followed by more frequent questions.

Communication Modes in Training

Whether your delivery medium is the classroom or computer, you will rely on three primary modes to communicate your content and instructional methods. They are text, visuals, and audio. In instructor-led settings, I recommend providing a workbook that will use a combination of text and still visuals to communicate instructional methods. In e-learning I recommend a combination of visuals, audio, and text. Here I quickly summarize what research tells us about best practices for the use of text, visuals, and audio. For more details consult the resources listed at the end of the chapter.

1. *Use visuals.* Relevant visuals plus words have been proven to lead to better learning than words alone. Therefore you should create graphics throughout your training. Because e-learning relies heavily on the screen to communicate, you should make heavy use of relevant visuals to communicate lesson content. Note that I said relevant visuals. Because they are so powerful, visuals that are off target can actually defeat learning. Stick with simple visuals that are related to the instructional objectives.

2. *Explain visuals with audio.* When explaining a complex visual, learning is better when the words are provided in audio rather than in text or with a combination of audio and text. Therefore in e-learning, make audio your default option to explain on-screen graphics—both still and animated. In classroom instruction, the instructor should project relevant visuals on the screen and explain them via narration.

3. *Integrate words into visuals.* If your delivery medium does not offer audio capabilities, such as either a workbook or e-learning viewed by individuals lacking access to audio, you will need to explain visuals with text. Be sure to place the text as close to the visual as possible for best learning results.

4. *Avoid use of text **and** audio to explain visuals.* There is a common misconception that it's a good idea to explain a visual with both text and audio narration of that text. In e-learning you will hear narration of the on-screen text. In the classroom, the instructor will read content on slides or in workbooks. Research has proven that the use of two redundant modes leads to poorer learning than only one mode. Use either text or audio to explain a visual—not both.

5. *Keep it lean and simple.* Research has shown that extra content such as anecdotes added for entertainment purposes or technical details added to embellish the main content

will depress learning. Avoid extraneous words, visuals, or sounds that are not relevant to your instructional goal.

6. *Use structured writing.* For display of content and methods in workbooks, I recommend a structured writing format to keep consistency among lessons and to maximize learning efficiency. In the next section I discuss structured writing in detail.

Communication Guidelines for Instructor-Led Training

For instructor-led technical training—either in the face-to-face or virtual classroom—provide a workbook that summarizes most of the key content and displays the practice exercises to be included in the lesson. Research indicates that in instructor-led environments where the learner must absorb information at a rate controlled primarily by the instructor, mental load is high. To ask learners to engage in a secondary activity such as extensive note-taking to document content will overload memory and depress learning. For consistency and quality-control reasons, I recommend using a structured writing technique to display your content and practice exercises in your workbooks. In this section I discuss each of these recommendations.

Note-Taking and Learning

Because note-taking is such a pervasive activity during learning, a number of research studies have evaluated its effectiveness. In general, it appears that note-taking does improve learning when the learner has control over the rate of instruction. For example, when reading a text or reviewing an asynchronous e-learning course, taking notes helps build new knowledge structures in memory. In contrast, taking detailed notes in an instructor-paced environment typical of most face-to-face and virtual classrooms will more often than not cause distraction. Taking notes will also absorb much precious instructional time—time which could be better used to engage participants in practice opportunities. As a result, I recommend providing notes for all instructor-led training events.

Structured Notes and Learning

If you have never seen structured writing formats before, some of the layouts may look quite strange at first. For example, look at the introduction to the toothbrush lesson in Figure 2.2. Note the use of marginal labels and lines to separate text. These are some of the most visible differences you will note in the structured materials. These formats are not arbitrary in their design. Research on layout and display of text has shown these to be especially advantageous to speed access and retrieval of technical information.

Structured writing formats can speed access to information. For example, compare the information displayed in Figure 2.3A with that in 2.3B. Experiments have shown that information is retrieved faster when a structured writing format, such as that shown in 2.3A, is used to organize and display content. The reason is because the layout and signaling of information using white space, tables, and labels makes information access more efficient. When doing a timed demonstration in my seminars, I find a typical 1- to 2-minute average advantage in retrieval of information in a structured writing version. Multiply this by the amount of technical information buried in typical business documentation to appreciate the value of the structured writing methodologies.

Research on text design shows that the information retrieval benefits from:

- Effective organization of related chunks of information

- Identification of that information with headings and labels (While my examples use marginal labels, research has shown that embedded labels are equally effective.)

- Elimination of unnecessary words by use of tables and by lean text

- Use of white space to separate information

Figure 2.3. Structured (A) and Typical (B) Information Displays Text Design Research

A. Structured

Company Reorganization

Background Competition from global markets has made it necessary to be more efficient. We have decided to create a leaner, more efficient operational structure.

Changes

Name	Moves from	To
John Smith	VP, Research	Product Manager
Reshmi Gaant	Dir. Finance	CFO
Jed Richardson	VP Training	Dir. HR
Sarah Jones	Assistant VP Operations	VP Ops.

No Layoffs We anticipate that attrition alone will bring us to an optimal staffing level. No layoffs will be needed.

B. Typical

Memo to All Staff

As many of you know, we have faced a difficult year with increased competition from foreign suppliers whose labor costs are lower than ours. Overall we have had a reduction of 23% of market share due to lower pricing of our competitors' product lines.

Rather than doing massive layoffs, we decided to use this opportunity to evaluate our operating processes and to make strategic decisions on ways to operate in a leaner, more cost-effective mode.

Our reorganization will include four major changes in senior management. Our VP of research, John Smith will assume duties of product manager and his position will not be replaced. Reshmi Gaant will increase her responsibilities by assuming the open CFO position. We will consolidate our Human Resources operation which will be led by Jed Richardson. Sara Jones will expand her roles moving from Assistant VP to VP of operations.

We believe that we can meet our financial goals with no layoffs. Attrition alone should bring us to optimal staffing levels in 3 years.

Advantages of Structured Notes

In addition to improved readability, structured notes provide the following advantages:

1. *Consistency.* Your instructional events will gain three types of consistency. First there is the obvious consistency in the appearance of your materials. If there are multiple course developers, each lesson will have a consistent format. Likewise, if topics from one lesson can be repurposed into different lessons or courses, the swap is easy. The second type of consistency goes below the surface. In Chapters 3 through 7 I will summarize the key instructional methods associated with the major technical content types. The structured formats based on content types promote quality consistency by ensuring that each content type is supported by optimal instructional methods. The third consistency emerges during course delivery. By documenting the key content and practice exercises, the quality of the course will not rely as exclusively on the instructor as when there is minimal or no documentation. This ensures a high-quality and more reliable learning environment across multiple classroom sessions.

2. *Time savings.* Using a template approach will save development time. Following a standard methodology speeds up course design and development. In addition, having a set of well-documented notes will save time during the learning event compared to classes where learners are required to document the content through note-taking.

3. *Better learning environments.* Rather than devoting most instructional time to lectures, with the content already documented, instructors can summarize and illustrate it and then promote learning through practice. Instructors will move from information givers to learning facilitators.

4. *Reference guides.* When a formal learning event is over, participants are expected to apply their new skills back at the workplace. However, there is rarely sufficient time to practice all of the new skills during the training so that learners can perform without any reference support. Useful documentation provides learners with a reference that will enhance transfer of learning.

Tradeoffs to Structured Notes

Naturally it will take more time to develop structured student notes than to prepare a deck of slides. You will have to consider the cost-benefit of detailed notes in your instructional setting by evaluating your course development resources, the criticality of the skills to be trained, the longevity of your content across time and across learners, and the performance value of having written documentation. In some cases you may compromise by preparing minimal handouts that are scaled back to working aids. In other cases you may provide skeleton notes with page titles, topic labels, and key visuals. The learners will need to fill in the gaps. In still other circumstances you may provide detailed notes. In this book I will show how to create workbooks with detailed notes. You can then determine how to adapt this approach to your different instructional contexts.

By evaluating the results of your training through testing, you can demonstrate the effectiveness of whatever approach you use. It is ultimately a management issue to decide on the cost-benefit of more and less effective training techniques. But until someone evaluates the outcomes and presents alternative approaches, there is no basis to show the benefits of improved training techniques.

The different content types and associated information-layout techniques in this book were originally developed by Robert E. Horn in the mid-1960s. Horn's method, which is taught in Information Mapping® seminars, is a comprehensive set of tools and techniques for identifying what needs to be communicated, organizing and managing large amounts of complex information, and presenting the information in consistent formats supported by human-factors research. While Information Mapping focuses on documentation primarily for reference purposes, I have found the basic methodology works well to summarize your technical content for training purposes.

Slides and Workbooks During the Class

Once the workbook is developed, you will want to prepare slides that the instructor will use during the class. I recommend that you format slides to emphasize correspondence between the slides and the workbook. To do so, incorporate key visuals or text phrases in the workbook into the slides. Where the workbook includes a visual, reproduce that visual on the slides. Where there are no visuals, use key text phrases. On some slides you may want to reproduce important examples. On all slides, post the workbook page number to help participants synchronize the workbook and slides. In places where the workbook includes a practice exercise, the slide might simply display the word PRACTICE and refer participants to the correct page. You can see a sample workbook page from a lesson on graphics in Figure 2.4 and one of several slides the instructor used to discuss the content in Figure 2.5.

Slides as Handouts

In some cases, instructors use copies of the slides as their sole handout or incorporate copies of the slides in the workbook. For most technical training, I discourage using printed versions of slides as your sole handout. Well-designed slides emphasize visuals and use minimal text. The instructor fills in the details. Therefore your learners will need to take notes to capture those details, which will lead to distraction during the class. If you have ever reviewed slides you've received at a conference, you may have experienced a sense of loss as you tried to decipher your notes.

Alternatively, you may opt to include the slides in the workbook with content details summarized above or below the slide. If you plan to use this approach, select a workbook format that allows

Figure 2.4. Workbook Page on Types of Graphics

From Clark Training and Consulting

Reference Notes			Your Notes
Communication Functions of Visuals			
Function	**Description**	**When to Use**	
Decorative	For aesthetic purposes; to add humor	Use sparingly; avoid distractions of seductive details	
Representational	To show what something looks like	To illustrate forms, screens, objects relevant to your goal	
Mnemonic	To provide a memory cue	To recall factual information	
Organizational	To illustrate qualitative relationships among program concepts	To give previews, overviews among concepts	
Relational	To show quantitative relationships	To show relationships among numeric data	
Transformational	To show changes in time or space	To illustrate procedures and process changes	
Interpretive	To illustrate principles or abstract concepts	To build deeper understanding of content	

Figure 2.5. A Slide on Types of Graphics

From Clark Training and Consulting

Communication Functions of Visuals

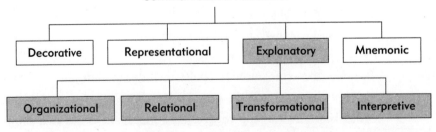

See Class Notes: Page 2-1-3

flexibility regarding placement of slides and text. A rigid format such as one slide per printed page will often be too constraining and/or wasteful. In some cases, a given topic may require only one slide with a key visual that will be extensively explained by the instructor. This topic might translate into a single slide and two or three pages of text. In other cases, a topic may require several slides each with a relatively smaller explanation. This topic would require several slides each with a few lines of text.

In my seminars I have found it most practical to build the workbook first, being sure to incorporate key visuals throughout. Then I build the slides based on the workbook, but do not provide separate handouts of the slides.

Communication Guidelines for Asynchronous e-Learning

Because the computer screen has different display properties and capabilities than a printed page, I recommend a greater reliance on visuals than on text. Visuals will promote greater engagement. Explain visuals with audio narration as the default option. However, allow learners to select a text alternative in the event that they do not have sound projection capabilities and/or have hearing impairments. However, whether you are teaching content with text, visuals, or audio, you will see in Section Two that the basic methods tied to content types will remain the same. You will simply implement those methods in different ways. I will show you how to do so in Chapters 3 through 7.

Many asynchronous e-learning courses also provide an adjunct workbook. The workbook serves as a summary of lesson content and may also include working aids to guide learners after completion of the training. Including a workbook will save time that learners may invest in taking notes during the lesson. It will also ensure an accurate repository of the content for later reference. If you plan to include a workbook, follow the formatting guidelines recommended in the previous section.

Educational Taxonomies and Instructional Methods

Chapters 3 through 7 will focus on the detailed instructional methods that fit into the lesson structure described in this chapter. In Chapter 1, I made the point that these methods are based on research guidelines from learning psychology. In this section I will introduce the *Content-Performance Matrix,* the classification system or taxonomy on which the methods in Chapters 3 through 7 are based. First I will give a brief historical background into the evolution and significance of educational taxonomies. If this background information is not of interest to you, bypass this section and go right to the overview of the Content-Performance Matrix.

Educational Taxonomies and Technology

Every significant technology, from medicine to engineering, has evolved through similar stages. They began as crafts, with designated practitioners. The barbers who served as medical practitioners in past centuries each had his own idiosyncratic approach to the craft. Before formalization or scientific principles, any one approach was as good as any other.

In the next stage, commonalties in problems or solutions are observed and grouped into classification systems known as taxonomies. In medicine, systematic recorded observations noted commonalties in symptoms, which were gradually classified as diseases. This stage symbolizes a major transition from craft to technology. Documented objective categories represent the beginning of the formalization of knowledge and the application of principles.

Bloom's Taxonomy

One of the first taxonomies of learning was presented by Benjamin Bloom (1956) in the 1950s in a model that included three major types of learning outcomes: *cognitive* (or intellectual), *affective* (or attitudinal), and *psychomotor* (or physical). This was the first time a formal systematic distinction was made among such diverse goals as learning how to ride a bike (psychomotor), how to solve a long-division problem (cognitive), and how to generate commitment to a democratic form of government (affective). Within each outcome different performance levels are specified from "lower-level" outcomes such as memorization of facts to "higher-level" outcomes such as analysis and synthesis of data. Many educational communities rely on Bloom's taxonomy today and you will hear references to "lower" and "higher" levels of learning.

A useful taxonomy will group learning goals to suggest different ways each should be taught. In medicine, a valid grouping of diseases categorizes symptoms according to effective treatments, because there are commonalties in underlying causes. In education, any powerful taxonomy would need to include categories that validly suggest differences in the ways the learning outcomes could be achieved through application of instructional methods.

Gagne's Conditions of Learning

To this end, Robert Gagne, in his 1985 book *The Conditions of Learning and Theory of Instruction,* introduced a more prescriptive taxonomy than Bloom's. He not only included five types of major different learning outcomes, but also described the *conditions* under which each is acquired. His work was one of the first that attempted to describe how instruction should be designed to provide the conditions that lead to different categories of learning. From the late 1950s into the 1970s, these conditions were expanded by a number of research studies that resulted in improved methodologies for teaching different types of content.

Merrill's Content-Performance Matrix

Building on the work of Gagne and the other research, M. David Merrill produced the taxonomy which, adapted somewhat, provides the basis for this book. I call this taxonomy the Content-Performance Matrix. I have found that the Content-Performance Matrix provides new instructional professionals with a very effective and succinct basis for designing effective instructional materials.

To learn more about any of these three taxonomies, refer to the sources at the end of the chapter.

An Overview of the Content-Performance Matrix

As shown in Figure 2.6, the matrix is two-dimensional with five types of content along the horizontal axis and two levels of performance along the vertical. Recall from Chapter 1 that two of the ingredients of instruction are the content and the learning outcomes. The content you teach can be classified as facts, concepts, processes, procedures, or principles. Your learning outcomes can be written at a remember or application level.

A remember outcome requires no deep mental processing, as it asks the trainee to recall or recognize information in an untransformed state. For example, you might ask your trainees to "list the steps needed to access customer records." To accomplish this task, the trainee memorizes the steps presented in the instructional

Figure 2.6. The Content-Performance Matrix: An Educational Taxonomy

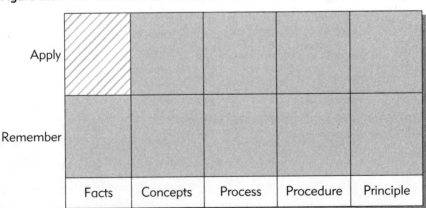

materials. However, being able to list the steps does not ensure that the trainee can actually access the customer records.

The bulk of your training should be targeted toward the application level, which requires the learner to apply the information as it will be used on the job. An application-level version of the above objective would ask your learners to "perform the steps to access the customer records." Your instruction should provide a job card or online support listing the steps and help the trainees to apply them as soon as possible. Detailed definitions of all five types of content and the two levels of performance will be presented in the following chapters.

How the Matrix Helps Develop Training

The power of the content-performance matrix rests in the matching of instructional methods to each of the cells in the matrix. For example, if you are teaching a concept such as "toothbrush" at the application level, there are certain informational displays and practice exercises you will need to use. An example of these particular displays is shown in Figure 1.5. If you need to teach factual information at the remember level, there are different guidelines to follow. The information displays are matched to the type of content involved.

Practice exercises are designed according to the level of performance. You will design a different practice exercise if you want your trainees to remember a concept, versus the one you would design if you want them to apply it. The practice exercise that matches the displays in Figure 1.5 would ask learners to pick out the toothbrush from other brushes found in bathrooms. In summary, the matrix provides the basis for rules you can use to develop training materials that work. This book is about those rules.

Chapters 3 through 7 define each of the five types of content and show you how to provide information displays and practice exercises—appropriate for print or computer delivery—at the remember and application levels. Although the *methods* will be the same for print and computer, the *formats* will change to accommodate the differences between 8½-by-11-inch pages and computer screens. As of this third edition, e-learning has evolved to include asynchronous self-study lessons delivered by CDs or by the Internet as well as synchronous instructor-led lessons. Thus I will use examples from both synchronous and asynchronous e-learning to illustrate computer delivery of the basic methods.

COMING NEXT

Instructional Methods Matched to Content Types

The key to effective technical training lies in the use of the right instructional methods to help learners achieve learning outcomes relevant to their jobs. These methods include information displays and design of practice exercises. This book will describe and illustrate the major instructional methods that research has proven most effective.

Chapters 3 through 7 present the information displays and practice exercises appropriate to each content type. Each chapter also discusses testing techniques you can use to assess learning. Chapter 3 focuses on one of the most common types of task: the procedure. Chapters 4 and 5 describe how to teach the most common forms of knowledge to support procedures: concepts and facts. Chapter 6 will describe processes that frequently provide job-related contextual information. Chapter 7 will present the second type of common task: principles.

For More Information

Anderson, L.W., & Krathwohl (Eds.). (2001). *A taxonomy for learning, teaching, and assessing: A revision of Bloom's taxonomy of educational objectives.* New York: Longman.

Bloom, B.S. (1956). *Taxonomy of educational objectives, Handbook I: The cognitive domain.* New York: David McKay.

Clark, R.C. (in press). *Building expertise* (3rd ed.). San Francisco, CA: Pfeiffer.

Gagne, R.M. (1985). *The conditions of learning and theory of instruction.* Austin, TX: Holt, Rinehart, and Winston.

Krathwohl, D.R., Bloom, B.S., & Bertram, B.M. (1973). *Taxonomy of educational objectives, the classification of educational goals. Handbook II: Affective domain.* New York: David McKay.

Merrill, D.M. (1983). Component display theory. In C.M. Reigeluth (Ed.), *Instructional design theories and models: An overview of their current status.* Mahwah, NJ: Lawrence Erlbaum Associates.

How to Teach Facts, Concepts, Processes, Procedures, and Principles

Chapter 3: How to Teach Procedures

Defines procedures as routine tasks and provides guidelines and examples of how to teach procedures in classroom and e-learning settings, as well as how to provide reference support for procedures in the workplace.

Chapter 4: How to Teach Concepts

Defines concepts as a critical type of supporting knowledge associated with learning to perform job tasks. Provides guidelines and examples of how to teach concepts in classroom and e-learning settings.

Chapter 5: How to Teach Facts

Defines facts as the other critical type of supporting knowledge in workforce learning training programs. Provides guidelines and examples of how to teach facts as well as how to provide factual reference resources in the classroom, in e-learning, and for performance support in the workplace.

Chapter 6: How to Teach Processes

Defines processes as important "how it works" content for workers involved with business, mechanical, or scientific systems. Provides guidelines and examples of how to teach processes in classroom and e-learning settings.

Chapter 7: How to Teach Principles

Defines principle-based or strategic tasks and provides guidelines on how to define the guidelines that underpin principle-based tasks as well as how to teach them in classroom and e-learning settings.

Figure 3.1. The Content Performance Matrix: Procedures

	Facts	Concepts	Process	Procedure	Principle
Apply				Perform the Procedure *Log on to the system*	
Remember				Remember the Steps *List the steps to log on to the system*	

3

How to Teach Procedures

CHAPTER OVERVIEW

The purpose of virtually all workforce training is to provide employees the skills they need to successfully perform job tasks aligned to organizational objectives. This means teaching tasks that are based on routine steps— tasks known as *procedures*. Additionally, you may be teaching how to perform non-routine tasks based on the application of guidelines. Chapter 7 will describe how to teach non-routine tasks known as principle-based tasks.

This chapter will show you the most efficient way to train procedures. After describing the types of procedures to be trained, I'll discuss how to promote learning at the *application* rather than the *remember* level. Then I'll illustrate how to provide the information displays and practice exercises that best teach procedural skills, both in print media for classroom instruction or working aids and in synchronous and asynchronous e-learning. Last, I'll summarize how to assess learning of procedural tasks.

What Is a Procedure?

A procedure is a series of clearly defined steps that result in achievement of a routine job task. Some typical examples include logging onto a computer, doing routine preventative maintenance on a pump, and taking a customer order. Procedures are done more or less the same way each time and can be clearly specified in a step-by-step format. A large proportion of all workforce training is procedurally based, focusing on helping workers perform the basic tasks to do their jobs efficiently, effectively, and safely. Teaching procedures effectively is therefore a critical skill for the training developer and instructor.

Linear and Decision Procedures

There are two basic types of procedures: *linear* and *decision*. Linear procedures are made up of clearly specified, observable steps, which are generally undertaken in the same sequence each time. You can document the procedure by watching an experienced employee do the task. Logging onto a computer and doing preventative maintenance are two examples of linear procedures.

Decision procedures are made up of two or more linear procedural sequences. A decision procedure is like a flow chart. The employee must make a decision that will lead her to continue to either sequence X or sequence Y. During the task analysis you have to find out the criteria of the decision by asking an experienced employee why he is doing X instead of Y.

How to establish credit for a new customer is a decision procedure. When the credit representative takes the order, he asks the customer a series of questions. Based on the customer's responses, the representative assigns a credit status code to the customer and may or may not require a deposit. The credit representative makes the decision based on a clearly specified set of company criteria. For example, if the customer owns his home or has an employment reference, a credit code of 3 might be assigned and the deposit waived.

Routine troubleshooting sequences are also based on a decision procedure. Depending on the symptoms of equipment failure, a technician might first try one of three tests to isolate the problem.

Depending on the results of these tests, a further series of tests would be conducted, followed by repair of the equipment.

Check Your Understanding

To be sure you can identify linear and decision procedures, try the short exercise in the Appendix under Exercise for Chapter 3.

Learning Procedures at the Remember and Application Levels

As mentioned in Chapter 2, the five types of content can be learned at two levels: remember and application. In order to apply a procedure the first time, the employee needs access to the steps. However, memorization of steps is generally a waste of time. You want the employee to *perform the procedure—not memorize steps*. Therefore, quickly move your instruction to the application level. In the next section I discuss and illustrate how to teach procedures so that your trainees quickly begin to apply their new skills.

Writing Procedure Learning Objectives at the Application Level

If your target lesson task is a procedure, you will write a lesson objective at the application level that requires the learner to perform the procedure. For example, the lesson objective for the tooth-brushing lesson would read: "You will be provided all needed equipment and will brush your teeth following the steps provided in the lesson materials so that you have no more than three dots on the red dye test." For the credit assessment task described above, the learning objective might read: "You will be given ten customer situations and you will input the deposit each customer must pay, with no errors."

Avoid writing remember-level objectives for procedures. For example, an objective that stated, "You will list the ten steps for brushing your teeth" *would not* be as effective as having the trainee actually perform the procedure.

Training Procedures

When learning procedures, employees need three basic instructional methods:

1. A clear statement of the steps that make up the procedure, with illustrations as appropriate

2. A follow-along *demonstration*

3. Hands-on practice with *explanatory feedback*

Whether you are training in a classroom or an e-learning environment, learning procedures depend primarily on having access to a documentation of the steps, seeing the steps correctly implemented in a demonstration, and having hands-on opportunities to practice the procedure with feedback. As you prepare to write up the procedural steps, take care to define them as discrete action steps, especially for audiences unfamiliar with the task. In fact, what might be a task for a novice audience (for example, "log onto the computer") could be a step for an experienced worker. It is always a good idea to test a rough draft of your documented steps with representatives from the intended learner audience. Start each step with an action word. Break long, complicated procedures into several smaller procedures. As a rule of thumb, your procedures should include no more than twelve to fifteen steps.

Teaching Procedures in the Classroom

In instructor-led face-to-face environments, learners will typically have a record of the steps in a training manual, view an instructor demonstration, and practice the steps in a hands-on mode using whatever equipment is involved in completing the procedure.

Formatting of Procedures in Manuals

Procedures are most cleanly presented in the training manual in *action* and *decision tables*. We'll look first at action tables and then at decision tables.

Figure 3.1 illustrates a procedure from the lesson on how to brush your teeth. The action table is divided into three columns: *step, action,* and *example.* Note that each step describes one simple, specific action. The newer the employee to the procedure, the smaller and more specific you need to make the steps. For example, new hires with little or no background in what you are teaching will require more detail than experienced employees who have some familiarity with the tasks.

Take a look at the entire page layout. A figure 3.1. page heading that states, "*How to . . .*" communicates that the information is

Figure 3.1. Action Table with Separate Example Column

How to Brush Your Teeth

Introduction — Now that you recognize the equipment you will need and we have discussed when you will need to brush your teeth, let's learn how to brush.

Step	Action	Example
1	Wet toothbrush with water from tap.	
2	Apply about ½ inch of toothpaste on bristles.	
3	Hold handle of brush and move bristles up and down against teeth.	
4	Open mouth and brush back teeth.	
5	Use glass of water to rinse out mouth. Repeat as many times as necessary until all the toothpaste is gone from your mouth.	
6	Wash off toothbrush and store equipment for next time.	

Demonstration — Your instructor will show you a videotape that illustrates the above steps.

Practice — In the lab area, you will find tooth brushing supplies to use as you brush your teeth. Your instructor is available to help you follow the correct procedure.

about performing a task. A brief introduction at the top of the page can either relate the procedure to the rest of the lesson or explain its importance. The Horn structured writing guidelines recommend marginal labels and lines to separate sections. However, white space and embedded labels are as effective. Figure 3.2 illustrates part of an action table for an Excel procedure, in which the illustrations are

Figure 3.2. Action Table with Embedded Graphics

How to Wrap Text in a Cell

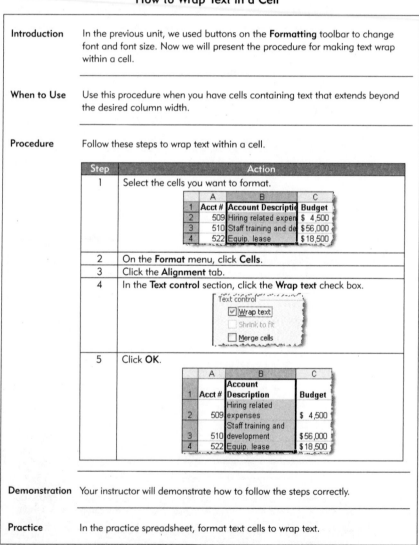

Introduction	In the previous unit, we used buttons on the **Formatting** toolbar to change font and font size. Now we will present the procedure for making text wrap within a cell.
When to Use	Use this procedure when you have cells containing text that extends beyond the desired column width.
Procedure	Follow these steps to wrap text within a cell.

Step	Action
1	Select the cells you want to format.
2	On the **Format** menu, click **Cells**.
3	Click the **Alignment** tab.
4	In the **Text control** section, click the **Wrap text** check box.
5	Click **OK**.

Demonstration	Your instructor will demonstrate how to follow the steps correctly.
Practice	In the practice spreadsheet, format text cells to wrap text.

embedded into the action steps. If some steps in your procedure will not have illustrations, consider embedding the illustrations into the cells with relevant steps rather than creating a third column, which will waste space with empty boxes.

If you are teaching a procedure with several steps that apply to a single screen or piece of equipment, consider displaying each step close to the relevant component of the screen or equipment as shown in Figure 3.3. Research has shown that more efficient learning results from integrated displays such as the one in Figure 3.3 than from displays that place words in a separate location from visual illustrations (Clark, Nguyen, and Sweller, 2006).

Decision Tables

Decision procedures are best documented using decision tables. Figure 3.4 includes a decision table from the tooth-brushing

---

Figure 3.3. Integrate Steps into the Visual

From Clark, Nguyen, and Sweller, 2006

How to Enter a Formula into Excel

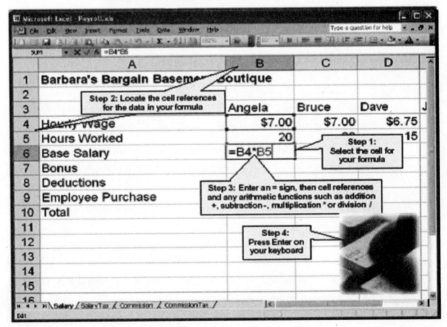

Figure 3.4. Three–Column Decision Table

When to Brush

The following table describes situations when you should brush your teeth.

If...	And...	Then...
You have just eaten	Your brush and paste are available	Brush your teeth
	Your brush and paste are NOT available	Rinse out mouth and brush later
You are going to bed	You have NOT brushed for four hours	Brush your teeth
	You have brushed in the last four hours	Wait until morning

lesson. Simple decision tables are written with an *If . . .* (condition statement) column followed by a *Then . . .* (action statement) column. They can be made a bit more complex, as in the "When to Brush" table, by adding the *And . . .* between the *If . . .* and *Then . . .* The page heading should summarize the decision being made, such as "*When to Brush*" or "*How to Resolve Display Station Problems.*" The short introduction explains why the decision needs to be made, and marginal labels help the reader quickly identify relevant sections.

Combining Action and Decision Tables

As mentioned above, decision procedures include two or more linear procedures. Sometimes it is most efficient to embed the decision table in the action table, particularly when the decision is relatively simple and directly tied into a larger action sequence. Figure 3.5 illustrates a procedure table with decision steps. Both Steps 1 and 2 include embedded decision steps.

Consider Flow Charts

If your target audience is familiar with flow charts, consider using them to document procedures with many embedded decisions as a more space-efficient alternative to decision tables. For example,

Figure 3.5. A Combined Action and Decision Table

How to Format a Financial Report

Introduction	Now that you have learned the Excel formatting techniques used by the Financial component, you are ready to apply those techniques to your monthly financial report.
Decision Table	Below is the sequence of questions to ask when formatting your monthly financial report.

Step	Action
1	Are all column and row titles in a bold font?

If	Action
No	• Select the cells you want to format • Click the **Bold** button (**B**)on the **Formatting** toolbar
Yes	Go to Step 2.

Step	Action
2	Are all financial numbers displayed with dollar signs and decimal points?

If	Action
No	• Select the cells you want to format • Click the **Currency Style** button ($)on the **Formatting** toolbar
Yes	Go to Step 3.

troubleshooting guides often use flow charts to document decisions that lead to one or more different actions.

Demonstrations of Procedures

Along with the steps of the procedure, the instructor needs to provide a follow-along demonstration to illustrate how to apply the steps. I recommend a follow-along demonstration to keep learners engaged rather than a strictly observational demonstration. To remind the instructor to provide the demonstration, consider referencing it at the bottom of the action table or flow chart in the student guide with a short sentence labeled "Demonstration." The sentence will read "Your instructor will now provide a demonstration of how to . . ." See the example at the bottom of Figure 3.1.

Support the demonstration in your instructor guide by listing all needed equipment and by providing any unique tips for conducting the demonstration. In some cases, demonstrations can be video recorded and shown in the classroom or embedded into e-learning. Video demonstrations are useful for procedures that are not practical to illustrate in the classroom setting, when consistency of the demonstration is a priority, or when the instructor is not a content expert. Be sure to keep demonstrations brief—especially when they are the observation type.

Teaching Procedures in e-Learning

One of the most common uses of e-learning is the teaching of computer application procedures such as how to use Microsoft Excel or how to enter a customer order into the corporate software program. Some e-learning is designed to be self-study. We call this *asynchronous e-learning*. Because asynchronous e-learning is designed to be used without support from an instructor, the appropriate use of instructional methods becomes even more critical. In other situations, procedures are taught *synchronously* in the virtual classroom. In Chapter 1, I stated that instructional methods determine instructional effectiveness, while the instructional media deliver the training. Trainees need the same information when learning procedures in asynchronous or synchronous forms of e-learning as in the classroom. They need clear statements of the steps involved in the procedure, a follow-along demonstration and hands-on practice.

Software tools make production of both asynchronous and synchronous e-learning software demonstrations easy. Some tools will capture the screens and the narration of an expert performing the procedure in a file format that can be integrated into asynchronous e-learning lessons. For synchronous e-learning, most virtual classroom tools include *application sharing* facilities allowing the instructor to project her desktop and perform software procedures visible to all attendees.

Consider these five issues when designing e-learning to teach procedures:

1. *Be Detailed.* For asynchronous forms of e-learning that will not have instructor support, it is especially critical to carefully include all required steps to ensure learner success.

2. *Be Brief.* e-Learning benefits from a much shorter learning cycle than classroom instruction. Sequence information and practice segments close together to hold attention and support learning. Present a few steps at a time and have learners follow along soon thereafter by trying out the steps in a simulation.

3. *Be Flexible.* e-Learning permits great flexibility to accommodate training content for varied levels of learner experience. I mentioned above that the task "How to Log On" may be a step to an experienced employee or a task to a novice. In asynchronous e-learning, both can be accommodated easily. If the "step" *Log onto the computer* is displayed as a link, a novice learner can click on the link to get detailed substeps as needed.

4. *Reformat.* The computer screen imposes different format constraints than the 8 1/2-by-11-inch page. I discuss specific issues around screen layouts in Chapter 9.

5. *Use Audio.* You may have access to audio and video media elements in design of your e-learning. Audio explanations of visuals lead to more efficient learning of procedures than text explanations. In both synchronous and asynchronous e-learning, use audio narration to describe a sequence of steps illustrated visually on the screen. In the virtual classroom lesson on Excel shown in Figure 3.6, the instructor uses application sharing to demonstrate computer procedures. The learners hear the instructor explaining the steps as they see the actions taken on the screen. You can find additional information about media modalities in Chapter 9.

Teaching Computer Procedures in e-Learning

Both synchronous and asynchronous e-learning are often used to teach computer applications. The lesson includes a demonstration followed by practice. In the virtual classroom, the instructor uses application sharing to provide the demonstration, as shown in Figure 3.6.

In the asynchronous e-learning course illustrated in Figure 3.7, each lesson focuses on a specific task related to use of a computer-controlled telephone-management system. Each lesson includes a short demonstration (*Show me how to do it*) followed by a simulation practice (*Let me try it*). In asynchronous e-learning, the demonstration is provided by narrated animated screen captures, as illustrated in Figure 3.8. This lesson focuses on how to transfer a call. The steps are narrated during the animated demonstration. This lesson achieves maximum efficiency by explaining visuals with audio narration.

Figure 3.6. Virtual Classroom Excel Demonstration Uses Application Sharing

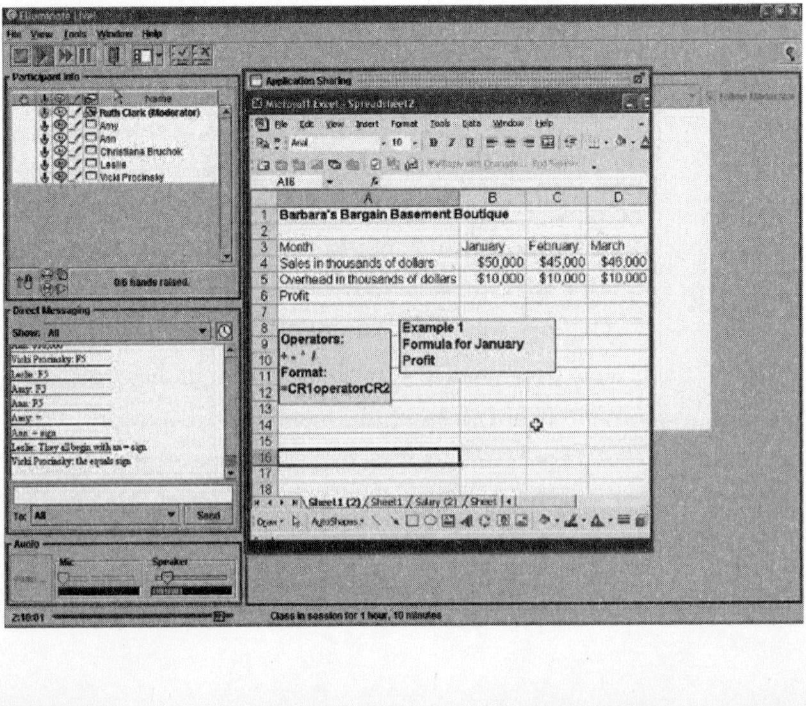

Figure 3.7. Menu for Asynchronous e-Lesson on Call Transfers

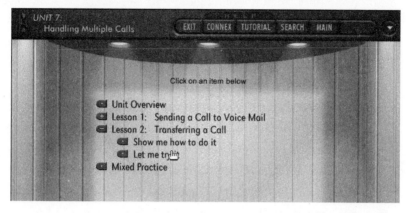

Figure 3.8. Asynchronous Demonstration of Call Transfer Procedure

Teaching Non-Computer Procedures in e-Learning

The computer is an ideal delivery tool for training computer-related procedures. Remember that, to learn a procedure, the trainee must perform it. When successful performance of a new task involves learning and accurately applying new motor skills, such as suturing a wound or assembling a drill, a delivery medium other than the computer should be considered, at least until virtual reality becomes practical reality for training purposes.

Sometimes e-learning can approximate real-life equipment actions that involve pressing buttons or turning valves, for example. In other cases, the learners already know the basic mechanics of the procedure. The focus of the lesson is on how to apply a new set of familiar actions in order to perform the task. In both cases, e-learning can potentially substitute for hands-on practice. In yet other situations, a preliminary e-lesson as part of a blended media solution can save time spent on actual equipment. For example, prior to attending a face-to-face hands-on lab session, learners take an e-lesson that familiarizes them with the equipment and the procedures. Then face-to-face time can be optimized by using it primarily for hands-on practice. Figure 3.9 shows an asynchronous multimedia lesson that uses audio narration and animated visuals to describe how brakes work. This type of lesson would be useful as a prerequisite for an instructor-led class on brakes maintenance and repair.

Figure 3.9. Asynchronous Lesson to Precede Instructor-Led Class
From Mayer, Mathias, and Wetzell, 2002

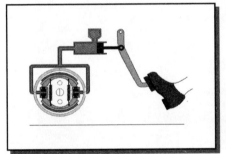 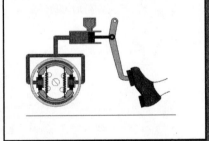

Practice Methods for Procedures

To be effective, training must provide relevant information interspersed with frequent practice exercises. Let's take a look at how to design practice that quickly gets the trainee performing the procedure. We know that practice is most effective when it is distributed throughout a learning event rather than assigned all at once.

Therefore, plan frequent and regular practice sessions throughout the lesson. Procedures can be learned at the remember or application level. Asking for remember performance in which trainees memorize steps is generally a waste of time. Instead, design hands-on practice opportunities that require the learner to perform the procedure in as close to job-realistic circumstances as possible.

Design of Classroom Practice

In the classroom, give the step summaries in action or decision tables in the training manual and provide a follow-along demonstration. Then assign short exercises that require the trainee to perform the procedure. Figure 3.10 summarizes some practice exercises designed for the classroom that require the trainee to practice the procedure with the instructor's help. Encourage the trainees to follow along on the action or decision tables as they begin. With more practice, they will refer less and less to this memory support. Any procedure that will be performed infrequently should be supported with

---

Figure 3.10. Sample Classroom Practice Assignments for Procedures in Classroom Training

I. Linear Procedure
How to Brush Your Teeth
Now that you have seen a demonstration of the correct way to brush your teeth, refer to the procedure table and use the materials in the lab to brush your teeth.

II. Linear Procedure
How to Process a Work Order
Work with a partner to complete the blank work order log and forms using the data in your student packet. Process each work order as demonstrated in class.

III. Decision Procedure
How to Determine Credit Status
Review the following dialogs between a credit representative and customer applicant. Use your working aid and write down the correct credit code that should be assigned.

reference guidance—either in text or embedded in the application. I describe reference guidance, also known as *performance support*, in more detail in the next section.

Design of e-Learning Practice

As I mentioned in the beginning of this chapter, e-learning has a short learning cycle in order to sustain attention. Thus, the learner needs frequent practice placed immediately after a demonstration. Figure 3.11 is taken from the practice segment of the call transfer lesson. In this segment of the simulation, the learner has made an incorrect response and receives feedback in the bottom text field accompanied an animated demonstration of the correct action. A common guideline in asynchronous e-learning is to follow the first mistake with an error message and hint and follow the second error with a demonstration of the correct response.

Figure 3.11. Practice Assignment in Asynchronous e-Learning

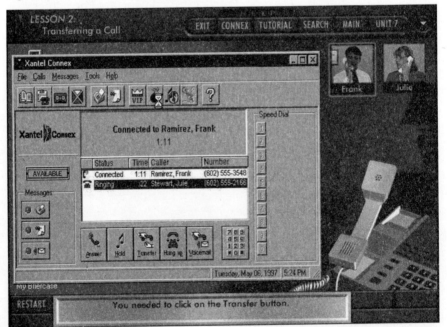

Performance Support for Procedures

Most workforce training programs lack sufficient time to ensure complete learning of the many procedures associated with jobs. Some important procedures may be thoroughly practiced during training. Others will only be introduced, and yet others may not be included in the training at all. Performance support in the form of procedural working aids—in either electronic or hard-copy formats are essential to ensure post-training task performance. For hard-copy aids, you can repurpose the action or decision tables created for the training manual. Electronic performance support systems—EPSS for short—are increasingly embedded in software. You are most likely familiar with the performance support features present in Microsoft Office applications such as the one shown in Figure 3.12.

If you are training procedures for which performance support has already been developed, incorporate that support into the training. For example, assign a case study and direct teams of two to complete the

Figure 3.12. A Performance Aid in PowerPoint

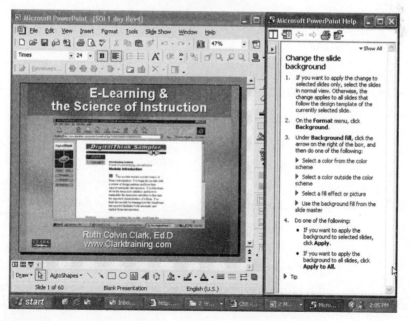

tasks by referencing the performance support. If performance support is not available, part of the training project should focus on developing suitable step-by-step performance aids. Be sure to organize these aids by the tasks that will be commonly performed on the job. For procedures conducted in a field location, you can put performance support on mobile devices, such as the example shown in Figure 3.13.

Figure 3.13. A Performance Aid on a Mobile Device

Evaluating Learning of Procedures

During the practice sessions, offer help as learners try out new steps. After training, you need to evaluate how effectively they have mastered the procedure without instructor help. The best way to do this for procedural learning is through a *performance* or hands-on test. Judge procedural learning by evaluating an end-product of the procedure or by direct observation of the performance. It is not valid to ask learners to list steps or unscramble steps, since these activities do not demonstrate application of the procedure. Evaluation of an end-product is easier than a performance observation. However, sometimes efficiency and safety issues dictate that the instructor observes learners performing the procedure.

Shrock and Coscarelli (2000) recommend the use of a yes-no checklist for evaluating procedures stating that "The checklist radically reduces the degree of subjective judgment required of the rater and thus reduces the errors associated with the observation" (p. 131). Figure 3.14 illustrates a checklist for a computer-related procedure. All trainees would be evaluated individually to determine that they could perform the procedure adequately.

Figure 3.14. A Checklist for a Computer Performance Test

Action	Yes	No	Comments
1. Log On Did the learner log on in one try in less than 30 seconds?			
2. Main Screen Did the learner access the PRD main menu in three keystrokes?			
3. Functions Did the learner start by using PRD functions to display datasets under their ID?			
4. Naming of data set Did the learner allocate a new fixed block data set using: Name 'ID>TRAIN>DATA 5 tracks primary space 3 tracks second space record length 250 bytes			

If a number of instructors will be administering a performance test, train them to consistently use the scoring guides when rating students. Gather all the raters together and have them independently score the same performance. Check the scores for consistency. Discuss discrepancies. Repeat this process for several sessions until scores are consistent each time. Technical methods for demonstrating reliability in scoring of performance tests are described by Shrock and Coscarelli in Chapter 15 of *Criterion-Referenced Test Development*.

How stringently you evaluate trainee performance depends on the criticality of the procedure. If the job requires near-perfect performance from the start, you will require a high score on your performance test, which means you will have to invest more training time. Compare these two training environments. A major semiconductor manufacturer uses equipment that must be operated very carefully to avoid expensive product and equipment damage. They invest considerable training time and use a stringent performance test before letting the trainee operate the equipment. In contrast, a customer service department teaches operators how to process customer orders over the telephone. They require a less stringent level of proficiency than the semiconductor manufacturing plant before allowing trainees to take customer orders. Keep in mind also that it is generally impossible to classify an individual as a master or non-master of a procedure on the basis of a single performance trial. Often three or four performance trials will be needed.

If the use of performance support is acceptable on the job, then access to performance aids should be available during the test.

To evaluate performance in e-learning, develop a simulation test similar to the simulation practice exercises that can be scored. Alternatively, schedule an instructor-led follow-up class where questions are asked and a performance test is given to assess proficiency. As a third option, require the trainee to produce some end-product, such as a budget spreadsheet in an Excel class, and transmit the result to a tutor.

For more information on construction, administration, and scoring of performance tests see *Criterion-Referenced Test Development* (Shrock & Coscarelli, 2000).

COMING NEXT

How to Teach Supporting Knowledge

In this chapter I provided guidelines for training routine tasks called procedures. There are always new terms and factual information associated with any procedure, and these should be included in the

"Supporting Knowledge" section of your lesson. The next two chapters describe how to teach two types of supporting knowledge that accompany procedures: concepts and facts.

For More Information

Clark, R.C., Nguyen, F., & Sweller, J. (2006). *Efficiency in learning.* San Francisco, CA: Pfeiffer.

Shrock, S., & Coscarelli, W. (2000). *Criterion-referenced test development* (2nd ed.). Silver Spring, MD: International Society for Performance Improvement.

Figure 4.I. The Content-Performance Matrix: Concepts

	Facts	Concepts	Process	Procedure	Principle
Apply		Classify New Examples *Select the valid signatures*		Perform the Procedure *Log on to the system*	
Remember		Remember the Definition *List the features of a valid signature*		Remember the Steps *List the steps to log on to the system*	

4

How to Teach Concepts

CHAPTER OVERVIEW

In Chapter 3, I described how to train job procedures. While training procedures you will usually make reference to equipment, tools, and technical terminology related to the steps of the procedure. For example, suppose Step 5 in your procedure is: "Place the zygometer in the reactive receptacle of the erylitizer when it reaches the suctorial stage." Many of the technical terms in this step need to be explicitly identified and included as part of the instruction. If you were teaching Step 5 and never explained zygometers, reactive receptacles, erylitizers, or suctorial stages, your trainees would be unable to perform the step, not because the procedure is difficult, but because the supporting information has been left out. As a result, the trainees may feel stupid or frustrated. That's why it's so important for you to systematically identify and teach all supporting information.

There are three major types of supporting information: concepts, facts, and processes. In this chapter I will describe how to train *concepts*. After defining concepts and showing some examples, I'll describe how concepts are learned at the remember and application levels. Informational formats for teaching concepts will be illustrated for both print media and e-learning. These will be followed by guidelines and examples for the design of practice exercises. As with all types of content, I will emphasize teaching at the application rather than the remember level. Finally, I will discuss how to assess learning of important concepts at the application level.

What Is a Concept?

Our everyday language includes many concepts, such as chair, mammal, house, and woman. Many of the sentence elements that we call "nouns" are in fact concepts. A concept is a mental representation or prototype of objects or ideas that include multiple specific examples. All of the concepts above represent a general class of "things," containing many examples. For example, the concept "chair" includes rocking chairs, folding chairs, and wheelchairs. The ability to communicate with concepts makes our language and mental representations very efficient. Suppose you had to hold a separate mental representation and use a separate word for every chair or house. Your memory would be cluttered by information that would take up resources more productively devoted to processing information.

All concepts have critical features or characteristics, and irrelevant features. The critical features are always associated with the particular concept; the irrelevant features vary from specific example to example. For example, all chairs share about four critical features. Nearly all chairs are intended for a single sitter and have a seat, a back, and some support from the floor to the seat. What are some of the irrelevant, or varying, features of chairs? Color, presence of arms, and type of support from the floor (typically legs but occasionally a podium, a rocker, or wheels) are just a few.

As illustrated in Figure 4.1, your mental representation of "chair" has abstracted the critical features, allowing you to recognize many examples of chairs that vary on the irrelevant features. Even if you were to see a type of chair unfamiliar to you, in a foreign country, for example, you would probably recognize it as a chair, based on the core features you hold in your mental representation. This is what makes concepts much more efficient than facts, which, as we will see in the next chapter, must be individually and uniquely held in memory.

Types of Concepts

It will be helpful in the next section on teaching of concepts to distinguish between two basic classes of concepts: *concrete* and *abstract*.

Figure 4.1. Concepts Share Critical Features and Vary on Irrelevant
Features

Concrete concepts have defined parts and boundaries that you can draw and label. "Bicycle," "house," and "chair" are examples. Abstract concepts are less tangible and cannot be directly represented using graphics. Examples of abstract concepts include "integrity," "credit," "deposit," and "concept."

Identifying Technical Concepts

If you are a technical expert, identifying the relevant concepts in your domain will be a challenge. That's because you are so familiar with them that you tend to forget that your learners are not. A good way to identify your concepts is to examine the steps you have documented in your procedures. Be alert for any terminology unfamiliar to your learners that would qualify as a concept. For example, suppose the tooth-brushing procedure illustrated in Figure 3.1 was for a learner from outer space. In talking to the alien, the instructor discovers the alien is familiar with mouths, teeth, and water. Therefore, there are two major concepts the alien doesn't know related to the procedure to be trained: toothbrush and toothpaste.

One good way to identify your technical concepts is to refer to the steps you have listed in your action or decision tables. Each new concept in the procedure should be trained in your lesson. Before I describe how to train concepts, let's take a look at how concepts are processed psychologically at the remember and application levels of learning.

Check Your Understanding

To be sure you can identify concepts, try the short exercise under Chapter 4 in the Appendix.

Learning Concepts at the Remember and Application Levels

Recall that all of the content types except facts can be psychologically processed at a remember and at an application level. At the remember level, the employee can recall the major critical features of the concept. For example, she might say: "A chair is a type of furniture intended for one sitter that has a seat, back, and support from the floor." Basically she is stating a definition that lists the critical features of the concept. However, as we see in Figure 4.2, just because a learner can state a definition does not mean that she can actually recognize an instance of the concept when she sees it. The real reason for teaching concepts in workforce learning is to help employees identify the tools or technical terms they will be using in their jobs. The ability to distinguish a concept is called *discrimination*. In everyday terms we would say, "Do they know one

---

Figure 4.2. Being Able to Remember the Features of a Concept Does Not Mean the Learner Will Be Able to Identify It

when they see one?" This skill will result from learning to mentally process concepts at the application level.

At the application level, the employee can identify or discriminate the concept by picking a valid example from a number of similar items. For example, you might give your learners a picture of ten items of furniture and ask them to circle all the chairs. Correct selections demonstrate that they have assimilated the critical features successfully so they can identify specific instances of the concept—even instances not previously seen.

In some cases, concept discrimination is the primary skill required by the job. For example, workers may be required to distinguish defective parts from those meeting quality standards. Or a loan agent may need to distinguish customers likely to repay loans from those who are not likely to do so. Transportation security personnel need to be able to discriminate potentially harmful objects from benign objects. In other situations, concept discrimination is a supporting knowledge linked to a larger procedure. If one of your action steps is to place the zygometer on the reactive receptacle of the erylitizer during the suctorial stage, the employee would need to be able to identify zygometers, erylitizers, reactive receptacles, and suctorial stages in conjunction with performing the task.

Writing Concept Learning Objectives at the Application Level

If the supporting knowledge section of your lesson includes a few critical concepts, you will need to write a supporting learning objective to establish a learning goal for this section. Typically, concept learning objectives are secondary to the major or terminal objective in a procedural lesson. For example, note the two lesson objectives written on the toothbrush lesson introductory page in Figure 2.2. The terminal objective requires the learner to brush his teeth correctly—a procedure. But the supporting objective asks the learner to identify toothbrush and toothpaste—two concepts. In some cases, a terminal lesson objective may focus on concept discrimination. For example, "Given a product run, the learner will identify all defective samples with no errors."

Write your concept objective at the application level by using an action verb that involves a classification activity. In the toothbrush lesson the learner will be given typical bathroom supplies and asked to pick out instances of the target concepts—the toothbrush and toothpaste—from the rest. The learner is classifying the target concepts by picking them out. Let's take a look now at how to teach concepts efficiently so workers will achieve the learning objective of making the critical concept discriminations.

Training Concepts

Instructional methods include giving required information in formatted displays and providing relevant practice exercises with feedback. To evaluate success, you need to test to be sure your objectives have been achieved. I'll first discuss the information requirements for teaching concepts and illustrate how you would format this information in a workbook and in e-learning. Then I'll describe how to design practice to support application-level learning, as well as how to test to assess concept learning.

Years of instructional research have identified the critical information required to successfully teach concepts. When teaching concepts, you must always provide a *definition* and *examples* of the concept. If possible, you can also provide *non-examples* and *analogies*. Let's look at each of these methods.

The Definition

A definition is a statement of the critical features associated with a concept. Writing a good definition may be more difficult than you realize, especially if you are so familiar with the concept that you need to really think about the key features. Our definition for chair might be: "A type of furniture intended for a single sitter which includes a seat, back, and floor support." Or suppose you were teaching the concept *nerd*. The definition might read: "A male student type who is somewhat socially awkward and typically can be identified by three or more of the following features: PDAs in their pockets and memory sticks around their necks, high-water pants, glasses, open-collar shirts, white socks, cowlicks, and skinny body type."

The Examples

An example is an instance of the concept. Once you have presented the definition, you make it concrete by presenting examples. If the concept is very simple, a couple of examples might suffice. However, if the concept is complex, with many features, provide several examples, each of which contain all the critical features and in which irrelevant features are systematically varied one at a time. As a general rule, start off with a typical example and move to less common illustrations.

Suppose you were teaching the concrete concept "dog" to a young child. After a general definition, you first show a picture of a "typical" dog. You might decide on a German shepherd. Your next example should vary only one or two of the irrelevant features. You might use a collie, where you have varied hair length and color. Your next example might be a sheltie, where you have varied size. You might also include a poodle and perhaps a couple of more exotic varieties. Carefully choosing a range of examples is critical. As illustrated in Figure 4.3, the learner looks at these examples and abstracts the critical features from the irrelevant ones. The child

Figure 4.3. Seeing Varied Examples Allows the Learner to Abstract a Mental Prototype of the Concept

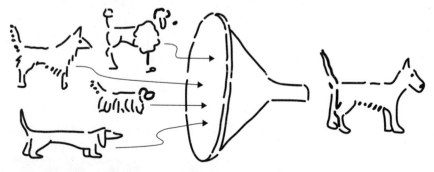

would note that dogs can have long, short, or curly hair; various colors, sizes, and body shape; but that all of them have the critical canine features, such as a muzzle, tail, etc.

Present examples of concrete concepts such as "dog" and "ery-litizer" with illustrations. If the concept has specific parts that are relevant, use a diagram with callouts to illustrate the concept. In contrast, for abstract concepts, construct examples in verbal formats. For example, if you are teaching the concept of good credit, an example might be: "Mrs. Jones has applied for service from our company. In checking her prior payment records, we find that she is recently divorced and did not have service in her name. But we discover she owns her own home and so can qualify for a good credit rating." During your task analysis phase, save lesson development time by collecting a number of examples from the work environment and modifying them to meet your instructional needs.

The Counter-Examples

When teaching concepts, a definition and examples are essential. It can also sometimes be helpful to add counter-examples. A counter-example is an instance of a closely related concept that could be confused with the lesson concept. Use counter-examples in situations in which Concept A is often confused with Concept B. After showing several examples, show the counter-example and explain why it is not an instance of the concept.

Note the instructor in Figure 4.4 using a chair as a non-example of a dog. Is this an effective counter-example? Probably not. That's because, other than the four legs, the two concepts share no common features. A better counter-example would be a concept whose features overlap those of the dog sufficiently to be potentially confusing. A cat is one obvious choice.

The Venn diagram in Figure 4.5 illustrates the function of a counter-example. Note that the counter-example B overlaps some of the features of the valid concept A and so can lead to confusion.

The effective counter-example stresses the ways the two concepts are distinct to minimize confusion.

Figure 4.4. Choose Appropriate Counter-Examples to Illustrate the Concept

Figure 4.5. Using Counter-Examples to Teach Concepts

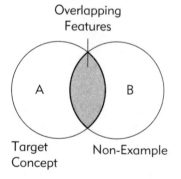

Analogies

In addition to counter-examples, analogies are another type of optional but often useful methods to teach concepts. An analogy is a representation that corresponds with a concept in function or form, but which is otherwise dissimilar. Analogies are typically drawn from a domain different from that of the material being trained. The pie analogy used to teach the concept of fractions is a classic example.

The teacher typically shows how the pie can be cut into halves and then into quarters, and so on. Analogies are efficient instructional techniques because they allow you to relate an unfamiliar concept to something the learner already knows, usually in a totally different domain of knowledge. While analogies are very powerful instructional tools, it is often difficult to think of an analogy that fits the concept you are teaching.

The key to an effective analogy is to find something familiar to your learners that accurately maps onto critical functions or features of your target concept. A misleading analogy is more damaging than no analogy at all. For example, a lesson teaching a credit rule that required the customers to have no more than two instances of poor payment used a baseball analogy of "*three strikes and you're out.*" This memorable analogy miscued learners because the actual rule stated "*two strikes and you're out.*" Knowing your learning audience is also essential to selecting an effective analogy. A technical analogy that works well for an engineering target audience might be meaningless for a less specialized population. An analogy that draws on U.S. cultural knowledge such as baseball may not work well for a global audience. If your target audience includes very diverse backgrounds, rely on everyday analogies that will be relevant across cultures. For example, rather than a baseball analogy, a soccer analogy might be more effective. Analogies can be presented in either pictures or verbal formats. Whenever possible, use pictures, as they can be more memorable than verbal descriptions.

I have presented the four basic informational types needed to teach concepts, that is, definitions and examples, and, when possible, counter-examples and analogies. Now let's take a look at how you might format this information, first in a training manual and then in e-learning.

Teaching Concepts in the Classroom

For training manuals, use the structured writing style introduced in Chapter 2. To introduce concepts, use a left-justified page heading

"What is X?" Then use marginal labels of "definition," "example," "counter-example or non-example" and "analogy" next to those elements on the page. Each section would be separated from the others by a line or white space. The goal in structured writing is to provide visible headings and to separate distinct bodies of information. Figure 4.6 illustrates these guidelines for the concrete concept of "toothbrush." The page heading "What Is a Toothbrush?" identifies the concept topic. Note the marginal labels and the example and counter-example which, because toothbrush is a concrete concept, are presented as graphics.

Figure 4.6. Information Displays for Concept Topic of *What Is a Toothbrush?*

What Is a Toothbrush?

Introduction	The most important tool you will need is the brush. Because most bathrooms contain several types of brushes, you need to identify the correct one.
Definition	A *toothbrush* is a small brush with a long handle, usually made of plastic, used exclusively for the purpose of brushing teeth.
Example	Here are pictures of two typical toothbrushes.

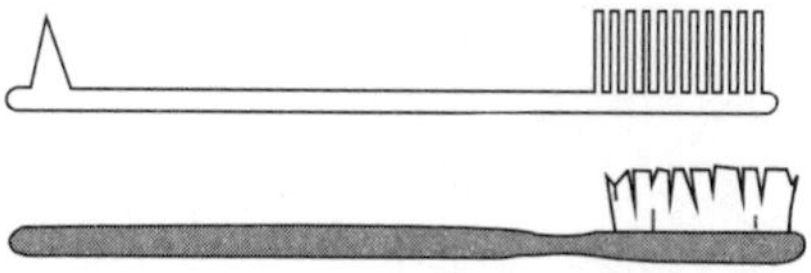

Non-Example	Here are pictures of two brushes that you cannot use to clean your teeth.

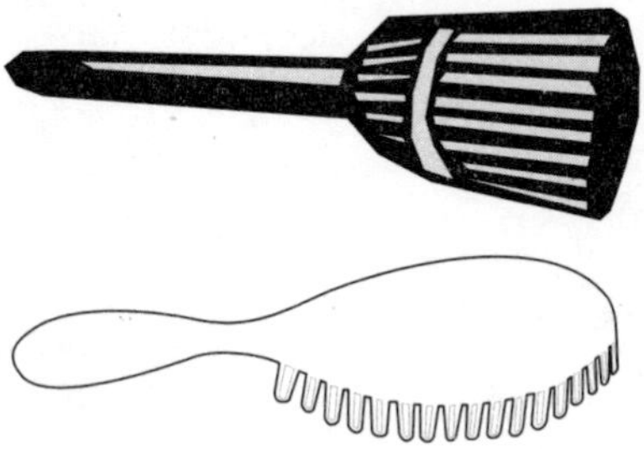

	They are much larger than a toothbrush and would not fit in your mouth.
Ownership	For sanitary reasons, each individual should own and use only his or her personal toothbrush.

For another example, take a look at the Nerd concept displayed in Figure 4.7. An introduction sets the context for the topic.

Figure 4.7. A Concept Display for Nerd

Credit: Tatiana Gill and Ann Kwinn

What Is a Nerd?

Introduction	So far in this course, you have learned about student and societal types such as the bully and the princess. In this lesson, you will learn about the nerd.
Definition	A nerd is a person with the following characteristics: • has more electronic devices than friends • wears clothing from another era, especially high-waisted pants • wears glasses (possibly repaired with tape or a safety pin) • exhibits poor hygiene and grooming • uses large quantitative words when possible • is a fan of comic books, video games, and science fiction, but not sports • has visions of being a superhero one day
Famous Examples	Napoleon Dynamite, Clark Kent, Urkel, the Nutty Professor, Bill Gates
Famous Non-Examples	• Queen Latifah – not a nerd because she wears stylish clothes • Johnny Depp – not a nerd because he grooms himself • Shaquille O'Neal – not a nerd because he likes sports • Superman – not a nerd because he really IS a superhero.
Analogy	A nerd is like a Volkswagon – slight of build, unassuming, and easily distinguished from more sophisticated models
Synonyms	Geek, Dork, Dweeb
Visual Representation	Characteristics of a Nerd: 1) electronic device 2) high-waisted pants 3) glasses 4) poor hygiene 5) large words 6) fan of comic books 7) visions of being a superhero

Credit: Tatiana Gill & Ann Kwinn

The definition is made especially clear by the use of bullets to emphasize the critical features. In this lesson the graphic example is reinforced by some famous examples, counter-examples, and an analogy. The graphic example uses a key-pointed table to reinforce the critical characteristics.

Finally, let's take a look at several examples from actual business training manuals. Figure 4.8 is from a lesson on how to write a legal and effective hiring interview. The concept drawn from that lesson is open versus close-ended questions.

Figure 4.8. Concept Topic from Lesson on Developing Hiring Interview Questions

What Are Open–Ended Questions?

Introduction	Now that you have explored behavior-based interviewing, let's look at open-ended questions.
Definition	*Open-ended questions* are those that cannot be answered by a single word or fact. They encourage the candidate to talk and provide the interviewer with valuable data on which to base a hiring decision.
Question Format	Questions do not have to • begin with the words what, where, when, who, why, or how *or* • end with a question mark
Examples	Here are three examples of open-ended questions. • How do you handle a customer's complaint call? • Explain how you handled a recent customer complaint call. • Describe a recent customer complaint call and how you handled it.
Non-Examples	Here are two questions that are not open-ended. Mary, have you ever used a personal computer? *This is not an open-ended question because Mary can respond with one word, yes or no.* John, what is your job title? *This is not an open-ended question because John can respond with a single fact, i.e., systems analyst.*

In some cases you will need to teach multiple closely linked concepts. Sometimes you can gain efficiency by teaching these together.
For example, Figure 4.9 teaches the difference between defects and
defectives for a quality control lesson. Note that the definitions
and examples are placed in a table. The examples for one concept
serve as counter-examples for the other. The visual provides a graphic
illustration that effectively depicts the differences between the two
concepts. Add visuals to illustrate concepts whenever feasible, as we
have good evidence that relevant visuals significantly improve learning (Clark & Mayer, 2007).

Figure 4.9. Teach Related Concepts Together

Defects vs. Defectives

Introduction	We are now at the next decision point in determining which control chart to use. Remember that this concept applies to attribute data only.
Definitions	Look at the definitions and examples below. Then turn to the next page for some practice with sample and defect concepts.

Terms	Definition	Examples
Defects	Number of faults per 1 item Fault(s) that cause an item to fail to meet specification requirements (there can be more than 1 defect per given part).	• 3 typing errors on a page • 15 holes on a roll of paper • Water tank with 10 leaks
Defectives	Number of items with 1 or more faults	• 1 defective page • 3 defective rolls • 5 defective water tanks

Leaking Water Tanks

Defects
Equal 5

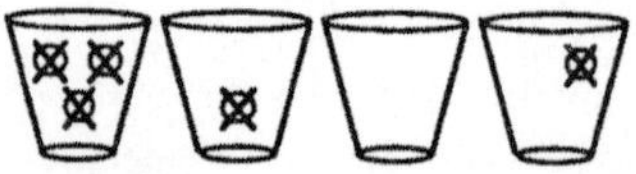

Defectives
Equal 3

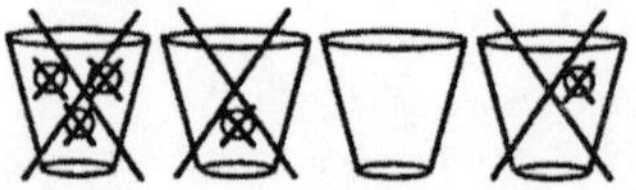

Now that we have seen a number of ways that the definitions, examples, counter-examples, and analogies can be formatted in a training manual, let's see how you would display similar information in e-learning.

Teaching Concepts in e-Learning

As mentioned previously, in design of e-learning the instructional methods linked to content types are the same as those used in print media. The major differences include layout of information on the screen and potential use of audio, color, and animation. Some important principles are keep screens uncluttered; use lean text and supplement with graphics and audio. Figure 4.10 shows one screen from an asynchronous lesson on formulas in Excel. The topic is

Figure 4.10. Excel Formula Concept Display in Asynchronous e-Learning.
With permission from Clark, Nguyen, and Sweller, 2006

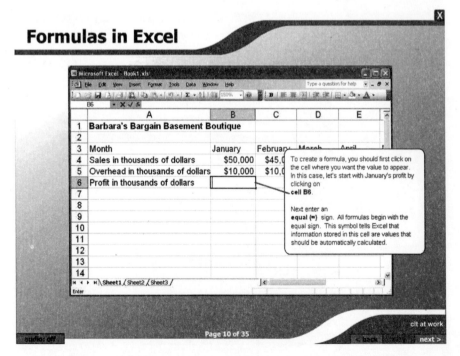

"What Is a Formula?" with emphasis on the formatting rules. The default option in this lesson is audio for greater learning efficiency. However, when audio is turned off, the explanation appears in on-screen text, as in this example.

Figure 4.11 is from a synchronous lesson on Excel formulas. The slide lists the key formatting features of a formula and includes two examples. The instructor explains the formatting rules using audio.

Figure 4.12 is drawn from an asynchronous e-learning lesson on creating websites. In this screen the topic of "text formatting" is explained with a definition, an example, and a counter-example. The user can click on either the example or counter-example to see an enlarged sample. Seeing the example and counter-example together with the definition helps the learner immediately compare and contrast features.

Figure 4.11. Excel Formula Concept Display in Synchronous e-Learning.
With permission from Clark and Kwinn, 2007

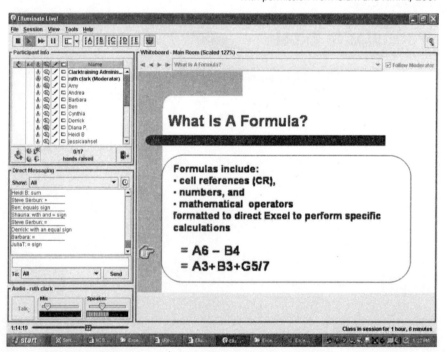

Figure 4.12. Screen Formatting Concept Display in Asynchronous e-Learning.

With permission from Element K and Clark and Lyons, 2004

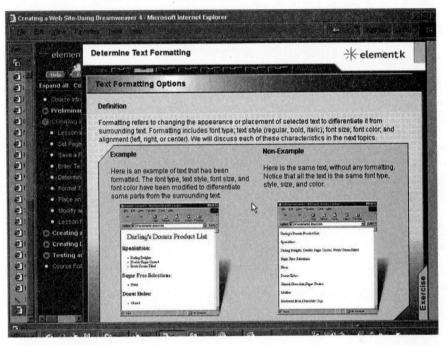

Once you have presented the major characteristics with examples and counter-examples, you will need to provide practice to ensure psychological processing of the information you have just given. Let's take a look at how to design practice exercises that accomplish your goal.

Practice Methods for Concepts

Recall that the major purpose of concept lessons is to help workers discriminate concepts so they can identify them on the job. For example, when working with zygometers, it's not as important that trainees can define a zygometer as it is that they know one when they see it—a skill called discrimination. Therefore, you will want to design practice to match your learning objective at the application level of performance.

To design practice at the application level, give the learner an assortment of new examples and counter-examples that were not used in the lesson and ask him to identify the valid examples. In our dog lesson for a child, we might present a page of animal pictures, asking the child to circle all the dogs. This is referred to as classification practice, because the trainee classifies each sample as a valid or invalid example of the concept. Successful classification indicates that the employee can discriminate effectively. This type of practice can be designed for both classroom and e-learning.

Design of Classroom Practice

In the classroom you can use a variety of paper-and-pencil formats or performance-type classification exercises. A performance exercise presents real objects that are classified by the learner. A performance concept practice is used when training inspectors. New inspectors must learn to discriminate between defective and non-defective parts. A performance exercise gives them ten parts to sort into "reject" and "accept" piles. A practice session on the concepts "toothbrush" and "toothpaste" might ask the trainee to pick out the brush and toothpaste from an assortment of common bathroom articles. If it is impractical to provide real objects, or if the concept is more abstract, use paper-and-pencil exercises.

Figure 4.13 illustrates the classification exercise that accompanies the topic: What Is an Open-Ended Question shown in Figure 4.8. To ensure understanding, the exercise asks for an explanation for each answer. The concept topic on defects and defectives shown in Figure 4.9 teaches two concepts simultaneously. The practice to accompany this exercise is shown in Figure 4.14. Here the learner reviews the visuals and states the number of defects and defectives. Paper-and-pencil practices can use a variety of formats, including circling the valid examples, matching, or multiple choice. The main point is to design the practice to get learners to classify the concept correctly to show that they know one when they see one.

Figure 4.13. Classification Exercise for Open-Ended Questions Topic

PRACTICE: Identify the Open-Ended Questions

Practice   For the questions listed below, write

- Y (yes) next to the open-ended question
- N (no) next to questions that are not open-ended

Explain why the question is or is not open-ended.

___ 1. Are you a professional electrical engineer?

Explanation _____

___ 2. How did you make the decision to apply for the cross-training program?

Explanation _____

___ 3. How long have you been a supervisor?

Explanation _____

___ 4. Describe how you handled a particularly difficult negotiation.

Explanation _____

Figure 4.14. Classification Exercise for Defect and Defectives Topic

PRACTICE: Defects vs. Defectives

Practice   In the following product lots, count the number of defects and defectives.

A.

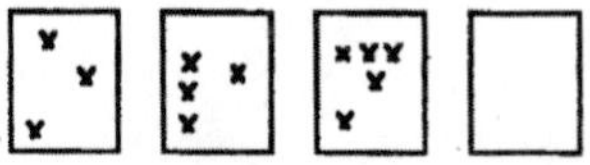

Defects: _____

Defectives: _____

B.

Defects: _____

Defectives: _____

Design of e-Learning Practice

To provide classification practice, e-learning can readily handle items such as multiple-choice and matching and simple on-screen text input. In asynchronous e-learning, you will need to decide on your item type as well as response options. Response options include: (1) click or touch the correct option; (2) drag and drop the correct option to a correct site on the screen; or (3) type in the correct answer. For example, in the exercise shown in Figure 4.15, the learner is asked to fill in the correct formula to achieve the calculation. The practice to support the concept of fields shown in Figure 4.16 uses a multiple-choice format. In this example, you see the feedback to an incorrect answer provided by the agent located in the lower-left-hand corner. If you are using audio to present your content, I recommend that you switch to text for exercise directions and for feedback. Audio is too transient and many learners will need to refer to the directions and feedback several times.

Synchronous e-learning likewise offers many facilities for classification practice, including chat, whiteboard activities, polling, and audio. The sample shown in Figure 4.17 asks learners to type in

Figure 4.15. Concept Practice on Excel Formulas in Asynchronous e-Learning
With permission of Clark, Nguyen, and Sweller, 2006

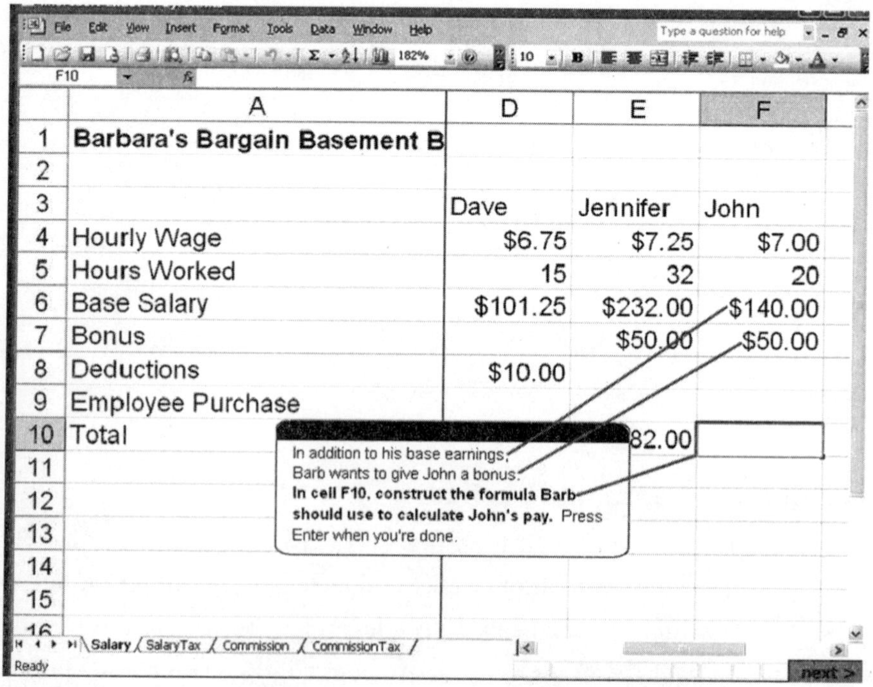

Figure 4.16. Concept Practice on Fields in Asynchronous e-Learning.

With permission from Clark and Mayer, 2007

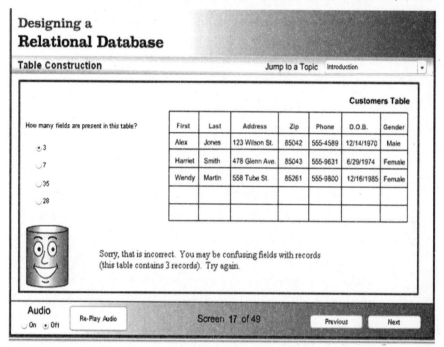

Figure 4.17. Concept Practice on Excel Formula in Synchronous e-Learning

With permission from Clark and Kwinn, 2007

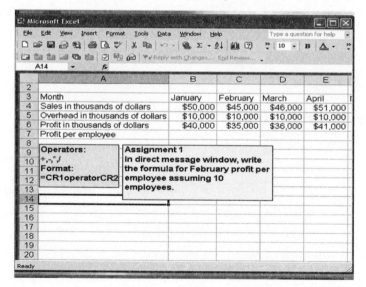

the correct Excel formula into the direct messaging (chat) window. Note the memory support window to the left of the assignment. To guide learners as they begin, the formula formatting rules are placed in this window.

Evaluating Learning of Concepts

Whether your training takes place in the classroom or in e-learning, you will want to be sure that your learning objective has been achieved. Can the learner successfully discriminate the new concept? If she can't, arrange for additional practice and assistance. To verify that concepts have been successfully acquired, create a test to match your learning objective, similar in format to the practice exercises shown above. If your objective stated that trainees would identify the toothbrush and toothpaste from a variety of bathroom supplies, both the practice and the test would provide supplies and ask learners to pick out the toothbrush and toothpaste. Your test items should use different examples from those used during practice sessions. No help should be provided during the testing period.

For detailed guidelines for constructing reliable and valid test items using multiple-choice or matching formats, consult *Criterion-Referenced Test Development* by Shrock and Coscarelli (2000).

COMING NEXT

How to Teach Facts

When you are teaching technical tasks, there is always supporting information that your trainees will need to learn in order to understand the steps or guidelines of the major lesson task. By following the guidelines in this chapter, you will be able to identify and train the concepts in your technical area. Chapter 5 will teach you to identify and instruct another common type of supporting information—facts.

Figure 5.I. The Content-Performance Matrix: Facts

	Facts	Concepts	Process	Procedure	Principle
Apply		Classify New Examples *Select the valid signatures*		Perform the Procedure *Log on to the system*	
Remember	Remember the Facts *Write out the password*	Remember the Definition *List the features of a valid signature*		Remember the Steps *List the steps to log on to the system*	

5

How to Teach Facts

CHAPTER OVERVIEW

In Chapter 4, I described how to teach concepts, one type of supporting information workers need to effectively perform their job tasks. Facts are a second category of supporting information. Factual information is very different from concepts in its characteristics and in how our memories process it. Teaching factual information requires instructional techniques different from those used for concepts.

In this chapter I'll describe how facts differ from concepts so you can identify factual information needed in your training program. We will then look at guidelines for the effective display of factual information in workbooks and on computer screens. Because facts can only be processed at the remember level, they are inefficient for human memory. We will discuss alternative ways to help trainees access the factual information they need for job performance, while reducing the load on human memory as much as possible. Design of practice exercises to bypass memory demands will be described for classroom and e-learning. Last, I will review techniques you can use to verify that your learners have successfully acquired (or can successfully access) the factual information they need to do their jobs.

What Is Factual Information?

Figure 5.1 illustrates the differences between concepts and facts. In Chapter 4, I described concepts as groups of objects or ideas that are given a common name, such as "chair," "mammal," or "house." Within each concept class are individual examples that share the common characteristics of the class but which also vary on a number of features irrelevant to the basic concept. For example, all chairs have a back and seat but may vary in color, material, presence of arms, etc. Unlike concepts, in which all members of the group share common properties, facts are unique, one-of-a-kind types of information. Specific data such as codes and passwords, unique interface screens, and forms are common examples of factual information.

Figure 5.1. Concepts Versus Facts: A Comparison

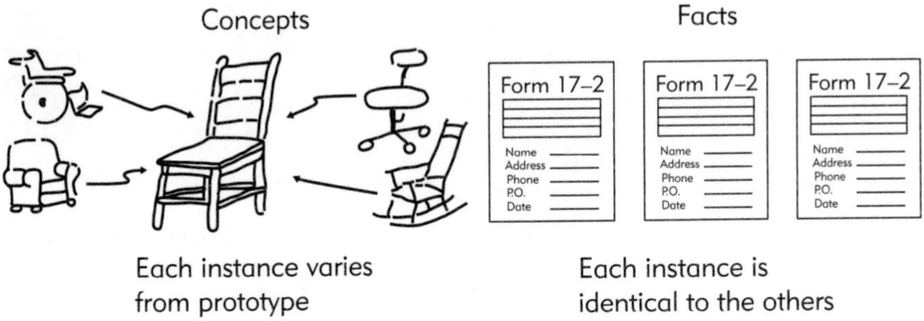

Concepts

Each instance varies
from prototype

Facts

Each instance is
identical to the others

Three Types of Facts

Factual information is commonly seen in one of three formats: concrete objects, unique data, and associations in statements. Frequently, factual information appears as concrete, unique instances of information such as specific forms, equipment, or computer screens where each instance is, for all practical purposes, identical to every other instance. The Tax Form 1040EZ is one example. While the term "tax form" is a concept because there are many different types of tax forms, the 1040EZ is a unique, specific form and would be considered a concrete fact. Concrete facts typically have parts and boundaries and often some identifying name such as the model UXB-700 Sharp fax machine.

A second common format for facts is specific data. Examples include dimensions and weight specifications of equipment, features and prices of products, numbers of employees per division, or names of division managers.

Third, factual information may appear in the form of statements that represent unique associations among concepts. For example, the statement "Our recent purchase order requested 534 chairs" would be an association between factual data of 534, and the concepts chair and purchase order. Other examples of fact statements could include: "The corporate assets total $45,000,000" or "The new vice president of marketing is John Jones." Note that all of these statements represent unique associations among concepts. For example, the last statement is an arbitrary linkage among the concepts of new, vice president, marketing, and a fact in the form of a name—John Jones.

Concepts are an efficient way to store knowledge. Once you have abstracted the common features of a concept such as chair, you will recognize objects you have never seen before as a type of chair. This is because you hold in memory a prototypic representation of the concept—an abstraction of "chairness." Unfortunately, facts must each be held in memory individually. This is because they have no common group features. By definition, a fact is a unique piece of information that must be individually held in memory to be known. Compared to concepts, facts are a much less efficient form of knowledge.

Identifying Factual Information in Job Tasks

The job tasks you are training always have related supporting information associated with them. Most of this supporting information is either conceptual or factual. Because technical experts are so familiar with the associated information, it is likely to have become unconscious knowledge. It is important to take a careful look at the steps or guidelines in your tasks to identify all technical concepts and facts your trainees will need to perform the task successfully. Because they are taught differently, you will need to distinguish between concepts and facts.

For example, look at the Excel procedure illustrated in Figure 3.3. In this lesson the learner is guided through the procedure of entering a formula into Excel to accomplish a given calculation. Step 3 asks the learner to format the formula using specific operators. For example, the correct operator for division is / and for multiplication the correct symbol is *. Only by knowing the specific symbols to use for the various mathematical operations can a functional formula be constructed. These specific symbols are examples of factual information.

When looking at your tasks to identify supporting information, ask yourself, "Are there multiple examples of this information that share common features but vary on irrelevant features?" If yes, you are dealing with a concept. Or "Is this a unique piece of information?" If yes, you are dealing with factual information.

Check Your Understanding

To be sure you can distinguish facts from concepts, try the short exercise in the Appendix under Exercise for Chapter 5.

Learning Facts at the Remember Level

If you refer to the matrix diagram on this chapter introduction page, you will note that all of the content types *except for facts* can be processed at the remember and application levels. Factual information is unique because it can only be memorized. This means that factual information can only be held "as is" in memory; it cannot be transformed, as can the other types of content. We all apply factual information every day in conjunction with the other types of content at the application level. But the facts themselves can only be held in memory in an untransformed state.

For example, in customer service training, representatives must learn to assign customers a credit code to classify their credit standing. In order to make the assignment, they must know the specific meaning of each unique code, which is factual information. When evaluating the credit application, the representative assigns a code based on the customer data provided and the meaning of the code. The worker must apply criteria to decide which code to assign. In fact, the action being taken by the representative is really a procedure, and the training would use a decision table, such as those shown in Chapter 3. In order to make the decision, the representative must have available the factual information regarding the meanings of the codes. Thus, although facts themselves can only be held in memory, they are used in conjunction with the other four content types, such as procedures, at the application level.

The formatting conventions of an Excel formula are another example of factual information that must be accessed during a procedure. While constructing a formula to accomplish a given calculation, the user must begin the formula with an equal sign. To achieve a specific calculation, the user must enter unique specific operators that correspond to add, subtract, multiply, and divide.

Because facts can only be processed at the remember level, they are very inefficient forms of information. As illustrated in Figure 5.2, that's because retrieval of information from human memory can be difficult. There is an unproductive tendency to require trainees to memorize too much information in many training programs. This is an unfortunate heritage from our earlier educational experiences, during which we devoted many hours to memory work. In workforce training our goal is to quickly build employees' confidence at performing their job tasks—not ask them to memorize a lot

---

Figure 5.2. Memory Storage of Concepts Versus Facts

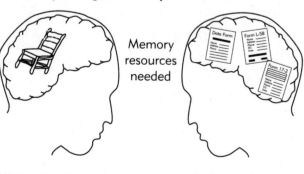

Memory
resources
needed

Efficient for Concepts   Inefficient for Facts

of information. Therefore, when I describe ways to train facts, I will emphasize techniques you can use to bypass memory limitations by providing external access to factual information.

Writing Learning Objectives for Factual Information

Because facts can be processed only at the remember level, objectives written for factual information are remember-level objectives. For example, the credit code lesson would have an objective such as: "You will state the meaning of each credit code." A lesson on Excel would include an objective such as: "For every calculation type, list the operator to perform that calculation."

However, stating the meanings of the operators as a separate objective is usually not necessary, and puts too much emphasis on memorization. Instead, write your objective at the lesson task level, which would require knowledge of the facts to complete. For the Excel lesson, your learning objective would read: "For the calculation scenarios, construct a formula to achieve each desired result." Here the objective is written for the task procedure, but its performance would require the trainee to have access to the factual information. Your goal is to minimize memorization of facts by having workers actively use those facts in conjunction with their job tasks. Your objective for the procedures in the credit code lesson might read: "You will be given ten customer application forms and will be asked to assign the correct code to each one." In order to achieve this objective, the learner would have to know (or have access to) the code meanings.

In conclusion, in general you can avoid writing separate learning objectives for factual information. Instead, incorporate the factual knowledge needed into the overall lesson task objective. There are some exceptions; situations in which rapid access to factual information requires memorization. In those cases, you may include a fact objective at the remember level.

Training Facts

As mentioned in Chapter 1, instructional methods include two major subsets: providing information and designing practice exercises for acquiring the content at the appropriate level of performance. An assessment is designed as follow-up to make sure that trainees have achieved the objective. In this section I will illustrate how to teach factual information, with formatting suggestions for print media and on-screen displays. Then I will discuss the design of practice for factual information, with an emphasis on ways to bypass unnecessary memorization or, when memorization is necessary, how to best support retrieval of facts.

Because facts are inefficient to hold in memory, I recommend that, when there are a number of facts needed to perform tasks or understand processes, you present them just prior to or in conjunction with the relevant step. This is an example of what is called "just-in-time" learning. Teaching a group of facts long before they will be used is generally a sub-optimal technique.

Training Facts in the Classroom

We saw that, to help learners build mental models of concepts, we provided definitions and examples. In contrast, facts are unique, and therefore each must be fully presented in an individual display. Depending on which type of fact you are training (concrete, specific data, or associative statement), apply one or more of the following formats.

Use Diagrams for Concrete Facts

Concrete facts such as specific forms, computer screens, or equipment are best presented via a diagram. In most cases, a simple line diagram is preferable to a photo to reduce mental load added by the extraneous details in a photo. If more descriptive information is needed, callouts can summarize parts or functions of the form, screen, or equipment. Figure 5.3 illustrates this type of display to summarize the names and functions of fax equipment. In conjunction with learning how to use any new equipment, a diagram that shows the major parts will serve as an orientation.

Descriptions or labels of graphic parts should be placed close by the relevant part of the graphic on the page, as in this example. Recent research shows that integrating descriptive text into the diagram reduces split attention that arises when learners have to mentally connect text separated from graphics (Clark, Nguyen, & Sweller, 2006). In some situations, however, you may have too much text to efficiently integrate it into a graphic. As an alternative place a table near the visual with the parts named and described, as shown in Figure 5.4.

Figure 5.3. Embedded Callouts Identify Parts of the Equipment

Sharp Fax Components   This is a picture of a Sharp fax with key components identified.

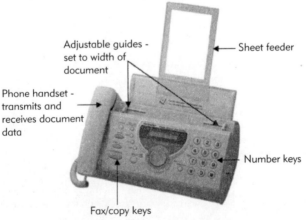

Adjustable guides - set to width of document

Sheet feeder

Phone handset - transmits and receives document data

Number keys

Fax/copy keys

Figure 5.4. Screen Parts Described with Callouts Linked to Graphic with Lines
With permission from Clark and Lyons, 2004

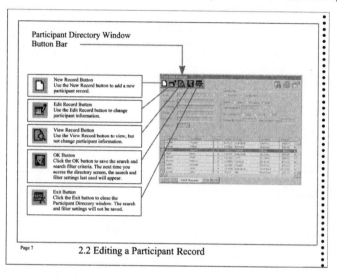

Use Tables and Lists for Data

Facts in the form of specific data can often be presented in tables or bulleted lists. Compare the back of the two seed packets illustrated in Figure 5.5. When sowing the seeds, the reference chart

Figure 5.5. Display Factual Data in Tables for Easy Access

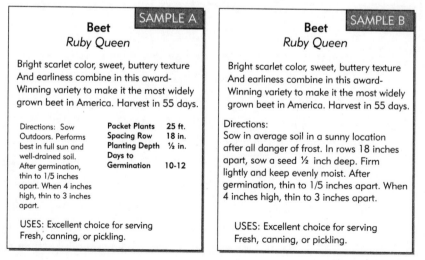

in Sample A is much easier to decipher than the text in Sample B. If you need to display a large amount of data, the tables or lists may consume entire pages or screens. On the other hand, a more limited set of facts can be integrated into other information displays.

Use Statements for Associative Facts

Facts that are unique associations among concepts are best presented in a sentence. For example, in our example concept topic *What Is a Toothbrush* in Figure 4.6, we included a sentence cued with the label "ownership" stating that "for sanitary reasons, each individual should own and use only his or her personal toothbrush."

Cueing the Reader with Unique Labels

Horn's research showed that, whereas the labels of definition and example can be universally used when presenting concepts, for factual information, you need to write a short descriptive label that signals the factual information to the reader. In Figure 4.6 the writer used a label of "ownership." Avoid use of the label "fact" because it does not effectively communicate the content of the factual information.

Teaching Factual Information in e-Learning

Factual information can be formatted for e-learning using many of the same displays used for manuals. Statements, tables, diagrams, and lists are all recommended. Figure 5.6 shows a table summarizing the main features of homo sapiens. The dynamic features of the computer allow some factual displays not easily achieved in paper. For example, different fact displays about homo sapiens can be accessed by clicking on the main topic tabs. In addition, the skull can be rotated allowing a 360-degree view. In Figure 5.3 we saw a print display presenting the parts and functions of a fax machine. The same information could be presented digitally using animation and audio. In a multimedia lesson the learner can click on any part of the graphic and hear a description of it and see it operate.

Figure 5.6. Facts About Evolution Displayed in Index-Card Graphic

Used with permission from TerraIncognita for ASU

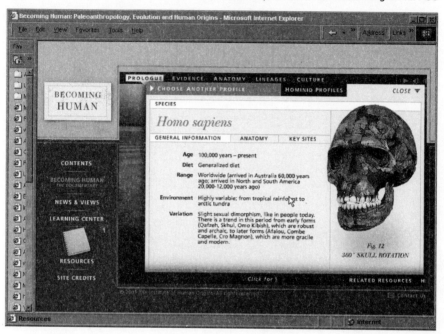

For reference purposes, databases allow dynamic displays of factual information on Internet sites. For example, most airline reservations systems, such as the one shown in Figure 5.7, use tables to display factual information related to flight choices.

Reference-Based Training

As the volume of factual information continues to proliferate, performance improvement will rely on workers who are good at accessing needed information rather than attempting to keep it in memory. For example, sales staff needs to keep updated on the latest product features. Or new hires want to check on their benefits package. Integrating reference and training is one way to build research skills in your learners. In *reference-based training,* factual information is displayed in reference documentation (either paper-based or online) while training is designed to "wrap around" the reference. For classroom training, imagine two documents—one (the training manual) for use during the training experience and the other (the reference

Figure 5.7. Airline Reservation Site Uses Table to Display Data

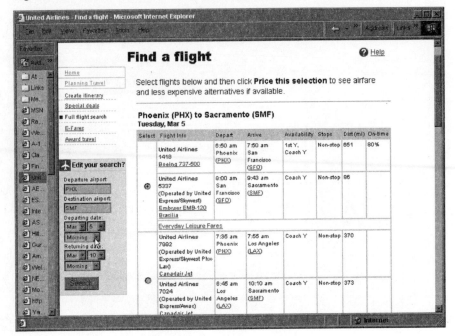

manual or websites) designed for use on the job. During training the learners are directed to the reference guides to look up procedural steps and facts needed to complete exercises. Exercises can be real-world scenarios or can use a "scavenger hunt" game format. The goal is to help learners become skilled in use of the reference resources during training so they use them effectively back on the job. For example, one large company created an intranet scavenger hunt game for new hires that required them to find important information such as benefits, contacts, travel coordination, and job-related resources such as sales proposal templates.

Use Inductive Learning to Engage Learners with Facts

Because facts can only be processed at the remember level, the teaching of factual information can quickly become boring and tedious. One technique to add engagement is to use an *inductive* question next to examples of the factual information as a way to create a learner-centered experience. Create an inductive learning experience by asking a question while showing the factual information needed to respond. For example, show an illustration of two or three sales tickets and ask, "What three items must be included on all sales tickets?" In Figure 5.8, I show a screen from a virtual classroom session on Excel formulas. The formatting conventions for Excel formulas represent important factual information needed to construct a workable spreadsheet. In this lesson the instructor shows two examples of Excel formulas and asks learners to use the chat window to type in the symbol used to start all Excel formulas.

Figure 5.8. Inductive Technique Used to Teach Excel Formatting Facts
With permission from Clark and Kwinn, 2007

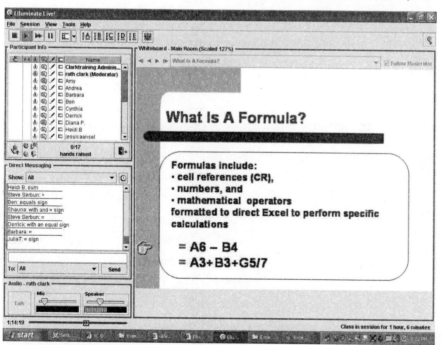

The topic of engagement brings us to the next major event of learning: design of practice exercises. Because facts can only be processed at the remember level, they present some unique challenges. The next section will describe practice techniques to use with factual information.

Practice Methods for Facts

When designing practice for factual information, first consider how workers will make use of the factual information on the job. Since facts can only be held in memory, assimilation of large numbers of facts is inefficient and tedious and should be avoided when facts can be provided in alternative routes. Four alternatives to practice facts include:

1. Incorporate factual information into practice of the linked task (procedure or principle) and add memory support job aids.

2. Use reference-based training models.

3. Provide drill and practice via e-learning to automate factual information when job aids are inappropriate.

4. Provide mnemonics as memory aids.

Practice Facts When Practicing Related Content

When learning to use Excel formulas, learners must start all formulas with an equal sign and use only the legal operators such as * for multiplication. After presenting the formatting rules, I recommend that you build formatting rules practice into practice of spreadsheet procedures. For example, in Figure 5.9 from a virtual classroom lesson on Excel, the instructor asks participants to type in the formula to achieve the posted calculation goal. To help learners recall the formatting rules, the instructor has placed a memory support window to the left of the assignment that summarizes the operators needed. This type of practice integrates factual information with the tasks

Figure 5.9. Virtual Classroom Practice of Excel Formula Formats

With permission from Clark and Kwinn, 2007

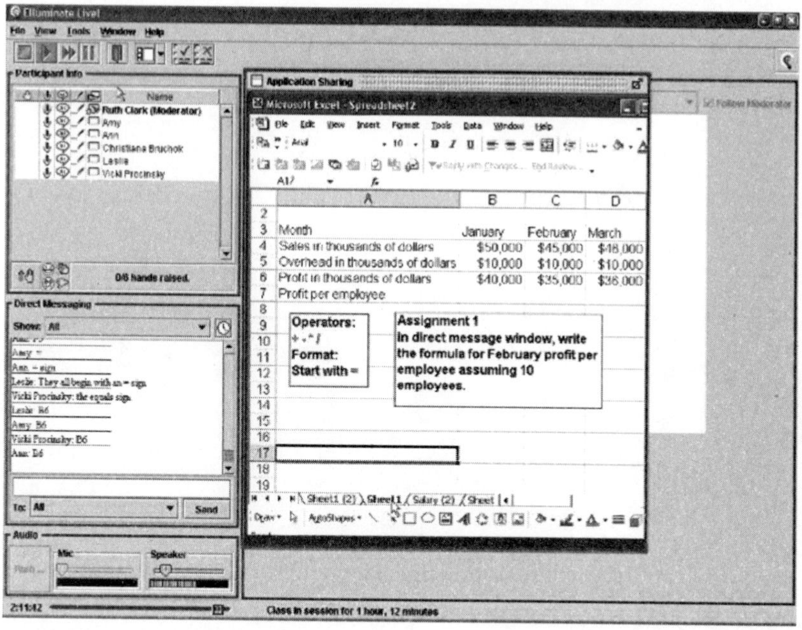

where the facts are used and is more interesting than asking learners to write out the operators.

A remember level practice that asks learners to label equipment parts is shown in Figure 5.10A. I recommend that in general you avoid this type of practice and instead assign a practice that asks learners to use facts as they would be used in the job. For example, the alternative exercise in Figure 5.10B asks learners to identify the functions of parts or to explain how parts would be used.

Use a Reference-Based Training Model

I described a referenced-based training approach in a previous paragraph. Start with the development of reference materials (either online or hard copy) and then develop your tutorials in ways that require the learner to use the reference to accomplish job-realistic assignments.

Figure 5.10. Two Practice Exercises on Factual Information

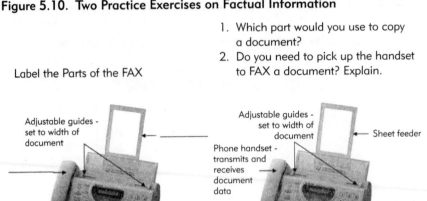

1. Which part would you use to copy a document?
2. Do you need to pick up the handset to FAX a document? Explain.

Label the Parts of the FAX

Adjustable guides - set to width of document

Adjustable guides - set to width of document

Sheet feeder

Phone handset - transmits and receives document data

Number keys

Fax/copy keys

Exercise A

Exercise B

Provide Drill and Practice to Automate Factual Information

Sometimes the use of a job aid is not acceptable due to safety or social conventions. For example, a train operator traveling at 120 miles per hour cannot take the time to access a job card to determine that the signal she is seeing requires her to slow down immediately. In these situations, the trainee must memorize the factual information in an automated fashion so the appropriate response is made instantly. Research shows that information can be automated after many hundreds of practice trials. Once automated, error rates AND response times decline rapidly. The computer provides a better tool for this purpose than the flash cards of earlier days. This is because the computer can simultaneously measure response accuracy and response time. Once accuracy is high and response times are very fast, the information has been automated. To make this kind of rote "drill and practice" more interesting, a gaming environment can be included that rewards accuracy and speed of response.

Provide Mnemonic Support Aids

Mnemonics have been used as memory support devices since the time of the ancient Greeks. An effective mnemonic provides an association between familiar and new knowledge so that the factual knowledge can be retrieved with the familiar. A common example is the *Every Good Boy Does Fine* mnemonic used by all new piano students.

In summary, when designing practice for factual information, integrate factual information into task practice. Provide memory support in the form of job aids or adjunct references. When access to memory resources during task performance is not practical or safe, design drill-and-practice exercises to help learners reach automaticity. Drill and practice is especially amenable to computer delivery, where both response time and accuracy can be tracked. In addition, providing an effective mnemonic will promote retrieval of factual information when a job aid is not desirable.

Evaluating Learning of Facts

Once you have designed some practice exercises that require learners to use the factual information needed to complete job tasks, you have a good format for design of test items to evaluate instructional effectiveness. Your test items, like your practice exercises, should be realistic simulations, using the factual information as it would be used on the job. In fact, when designing practice exercises, create some additional items similar in format to be used on your test.

Be sure that your test includes some items that make use of the job aid you have designed so you can verify that the trainees can effectively access the information provided on it.

In most workforce learning, we are focusing on accuracy of responses during training rather than on speed. However, in some situations, responses that are fast and accurate are needed from the start. In these cases, computer tests can accurately measure response time as well as response quality.

COMING NEXT

How to Teach Process Content

In Chapters 3 through 5, I have summarized the major techniques for teaching basic job procedures as well as related facts and concepts needed to perform tasks successfully. Sometimes, however, in order to complete a task, the employee needs to know about how something works. Knowing how things work can be essential to working effectively with or on those things. For example, doing maintenance on equipment may require knowing the overall functioning of the various parts of the equipment.

Any information that depicts how things work is called a process. In Chapter 6, I will distinguish processes from procedures and indicate when and how to teach them.

For More Information

Rossett, A., & Schafer, L. (2007). *Job aids and performance support.* San Francisco, CA: Pfeiffer.

Figure 6.I. The Content-Performance Matrix: Processes

	Facts	Concepts	Process	Procedure	Principle
Apply		Classify New Examples *Select the valid signatures*	Solve Problems Make Inferences *Given these symptoms, what could be wrong?*	Perform the Procedure *Log on to the system*	
Remember	Remember the Facts *Write out the password*	Remember the Definition *List the features of a valid signature*	**Remember the Stages** ***Describe how the Erylitzer works***	Remember the steps *List the steps to log on to the system*	

6

How to Teach Processes

CHAPTER OVERVIEW

In Chapters 3 through 5, I described techniques for identifying and teaching three of the core-content types that make up workforce training: procedures and their related concepts and facts. Often employees routinely complete their job tasks without understanding the bigger picture of which their work is a part. This lack of perspective can result in sub-optimal work performance. With the increased emphasis on knowledge work, the application of process knowledge is more critical now than it was in the past. In other situations workers are responsible for the maintenance or implementation of systems. To optimize their work, they need an understanding of the entire system. Systems are the focus of a type of content called processes.

In this chapter I will define processes, distinguishing them from procedures. We will then look at guidelines for effective teaching of processes, including some sample formats for workbooks and e-learning. Processes can be psychologically processed at the remember or application level. As with other types of content, I emphasize the application level of learning. I will describe practice exercises that encourage trainees to apply process knowledge *not just memorize it*. Last, I will provide some ideas for verifying that the trainee has acquired process knowledge at the application level.

What Is a Process?

While procedures are directive in nature, *processes* are descriptive. Procedures tell employees how to go about doing a task, while processes tell them how something works. For example, a lesson on how brakes work would be a process lesson useful to a mechanic trainee. A lesson on how Company X hires new employees might be useful to a clerk processing new hires or to a manager interviewing job applicants. A number of departments or jobs play roles in business processes such as hiring. An individual employee in personnel would follow his or her procedure which, along with procedures followed by employees in other departments, make up the hiring process. In short, process content focuses on how things work.

Three Types of Processes

Processes can be classified into three categories:

- *Business* systems depicting organizational work flows
- *Technical* systems depicting stages in mechanical systems
- *Scientific* systems depicting how natural phenomena occur

A business process consists of several stages performed by different employees or departments, resulting in achievement of an organizational goal. Employee hiring, customer billing, software development, and panel assembly are examples. These processes are combinations of individual tasks performed by different employees or functional areas of the organization. By contrast, a technical process consists of stages that involve equipment operations such as how a steam turbine works, or how glopples are manufactured in the erylitizer. Scientific processes focus on natural systems such as how a hurricane forms or how blood circulates.

Business processes are somewhat situational in that what may be a process in one setting could be a procedure in another. For example, in a moderate to large company, new employee hiring is typically a process involving several departments. However, in a small company, the hiring may be a procedure handled by a single individual.

When you do your job analysis at the start of your training planning, you will determine whether specific business functions are processes or procedures in the setting you are analyzing. Technical and scientific processes do not have this situational character. They will always be classified as processes since they describe systems such as equipment or weather that operate outside direct human control.

Motivational and Instructional Value of Processes

When an employee's work is part of an organizational system, an understanding of the process can support learning of job tasks, work motivation, and overall product quality. If, for example, the employee knows his internal customer—the recipient of his work product—is the accounting department, he will gain a more meaningful perspective of his role, along with the potential to define and improve work quality for his customers.

Often process knowledge contributes directly to a more meaningful application of procedures. If a programmer understands how the modules are interrelated for the system she is maintaining, her coding will be more effective than if she is working without that larger picture. If a technician is faced with a problem on some equipment, more effective troubleshooting will result from an understanding of the technical equipment process.

Identifying the Processes in Your Training

As you look at the tasks you are about to train, ask yourself whether there is an associated process linked to that task the knowledge of which would improve job performance. If you are training preventative maintenance of the erylitizer, it would be helpful to have a general understanding of how the erylitizer works, as summarized in Figure 6.1.

If you are teaching a new-hire to set up client investment plans on the computer using information from a variety of forms, you might provide an overview of the investment process, showing where the various forms originate and what happens to them after the new hire completes her part. Figure 6.2 illustrates this business process.

Figure 6.1. How the Erylitizer Processes Glopples

① Glopple Goo is added to intake chute.

② Heat reactor melts Goo at 450° C.

③ Liquid Goo purified in evaporative columns. Cools to 115° C.

④ Cool Goo enters die press.

⑤ Glopples soak 10 hours in acid bath.

⑥ Finished Glopples emerge.

Figure 6.2. How Automatic Investment Plans Are Processed

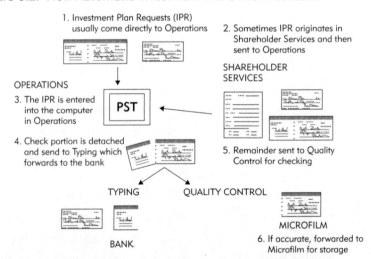

1. Investment Plan Requests (IPR) usually come directly to Operations

2. Sometimes IPR originates in Shareholder Services and then sent to Operations

SHAREHOLDER SERVICES

OPERATIONS

3. The IPR is entered into the computer in Operations

PST

4. Check portion is detached and send to Typing which forwards to the bank

5. Remainder sent to Quality Control for checking

TYPING

QUALITY CONTROL

BANK

MICROFILM

6. If accurate, forwarded to Microfilm for storage

In previous chapters I have used examples from a "How to Brush Your Teeth" lesson to illustrate techniques for teaching procedures, concepts, and facts. It might be useful to include a brief section on the process of tooth decay early in the lesson. This information should provide greater motivation for, and understanding of, the techniques for flossing and brushing. Figure 6.3 illustrates this natural system.

Processes can be presented briefly at a high level or more extensively with considerable detail. Part of your analysis needs to determine the amount of detail optimal for effective job performance. In many cases a short overview is sufficient and can be handled as a topic in a procedural lesson. However, for an extensive process for which considerable detail is needed, entire lessons or even courses may be appropriate. For example, a microchip manufacturing

Figure 6.3. How a Tooth Decays

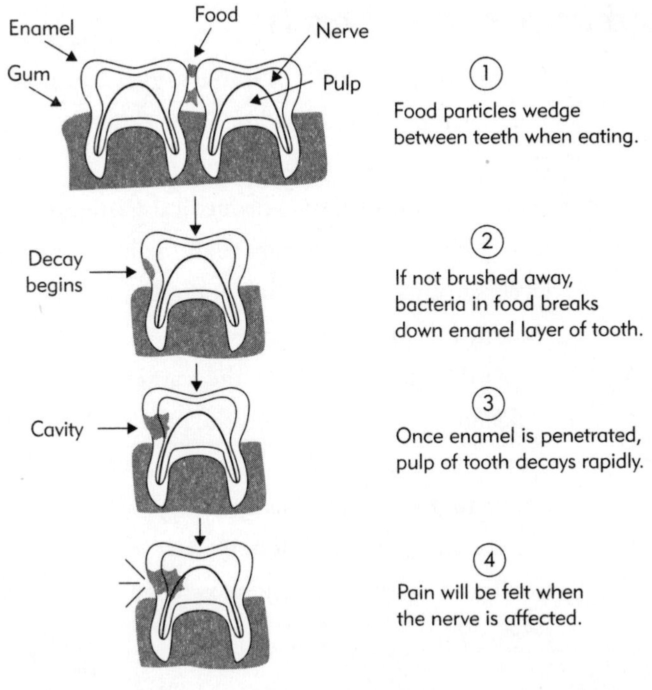

Enamel
Food
Nerve
Gum
Pulp

(1) Food particles wedge between teeth when eating.

Decay begins

(2) If not brushed away, bacteria in food breaks down enamel layer of tooth.

Cavity

(3) Once enamel is penetrated, pulp of tooth decays rapidly.

(4) Pain will be felt when the nerve is affected.

company developed an entire course on their equipment transfer process that described how new equipment was selected, evaluated, burned in, and installed at a beta site prior to wide-scale implementation.

A process may also form the overarching organizational framework for a course in which separate modules in the course include tasks related to a specific phase of the process. For example, my course on development of workforce training uses the instructional systems design (ISD) process as a framework. Lessons within the course focus on stages in the process, including needs assessment, job and task analysis, writing of learning objectives, and so forth.

Check Your Understanding

To be sure you can identify processes, try the short exercise in the Appendix under Exercise for Chapter 6.

Learning Processes at the Remember and Application Levels

Except for facts, all the content types can be processed at either the remember or application level. When learning processes at the remember level, the learner recalls or recognizes the basic stages associated with the process. The medical trainee might describe the process of blood circulation through the heart and lungs. In remember level performance, the information is essentially untransformed from the way it is presented in the lesson.

At the application level, the learner is able to solve a problem or make an inference based on the process. For example, the medical trainee might be asked: "What would happen if leakage between the left and right ventricle occurred?" or "What would be the effect of obstruction of the pulmonary artery?" To answer these questions, the learner needs to apply knowledge about the blood circulation process.

Application of process knowledge is especially important for employees involved in any form of troubleshooting for which they might have to resolve problems that occur in an overall business or technical process. Systematic problem solving in manufacturing operations begins by flow-charting the production process. Once the process is clearly defined, all employees associated with the process, from engineers to production workers, can brainstorm potential sources of the problem using cause-and-effect analysis.

Knowledge workers in particular are engaged with ill-defined tasks—tasks that are principle-based rather than procedural, as we will see in Chapter 7. Knowledge work inherently involves problem solving and adaptation to continuously changing circumstances. An understanding of the processes associated with that work is critical to making the continuous refinements required for effective work performance.

Writing Process Learning Objectives at the Application Level

Depending on the size and criticality of the process knowledge you have identified to include in your instruction, you may not always write a separate learning objective for the process part of your training. First ask yourself, "How critical is this process knowledge to the effective completion of job tasks?" If you feel that knowledge of the process is a "need to know," as opposed to a "nice to know," then establish an outcome goal for this knowledge by writing a learning objective. Otherwise you might include the process in your training but not hold your trainees accountable for the knowledge.

If you write a learning objective, start out by defining for yourself how the process is tied into the trainee's job. Write an objective that asks the learner to apply the process knowledge to solve a job-related problem. Avoid objectives that merely ask the learner to regurgitate the stages. Table 6.1 lists two sample learning objectives for process content.

Table 6.1. Sample Learning Objectives for Process Content

Audience	Objective
Erlitizer Engineer	You will be given descriptions of faulty glopples along with operator observations. Based on how the Erylitizer works, you will define possible problems in the manufacturing cycle.
Customer Service Representative	You will answer customer questions about credit-related documentation and direct customers to the correct department to resolve the situation.

Training Processes

The design of all effective instruction involves providing clear information, giving practice with feedback to help the learner assimilate the information, testing for achievement of the instructional objective, and repeating the teaching cycle as needed. Whether you deliver your training in the classroom or via e-learning, you will need to communicate the major stages of the process using words and diagrams. In situations in which the process knowledge is important to effective practice of job tasks, you will also add practice exercises that require learners to apply the process knowledge in job-relevant ways.

When the process you will present is complex, research has shown that learning is better when you precede the process explanation with any key facts or concepts associated with the process. For example, if teaching the process of how the Erylitizer works shown in Figure 6.1, learning would be better if you first presented the names and functions of the main parts of the equipment such as the heat reactor and the evaporative columns.

Teaching Processes in the Classroom

Like procedures, processes involve a series of events—stages rather than steps. Use either process tables or flow diagrams to present the stages.

Business processes usually involve information of a "who does what when" type. Figure 6.4 shows a process stage table that includes columns for stage, action, and responsibility. This example illustrates a process of preparing a monthly financial report that includes roles for the data steward, the assistant, and the analyst. A table for a technical process that summarizes how equipment works might include columns for stages, for example, *ignition* parts such as *ignition leads and spark plugs,* and description, for example, *When the ignition leads contact, the spark plugs fire igniting the fuel in the cylinder.*

A second method of displaying processes is to use a flow diagram either on its own or in conjunction with a descriptive table. Flow diagrams may be circular to illustrate ongoing flows, or linear for processes with defined starts and stops. Figure 6.5 illustrates a simple flow diagram for a technical process related to preparing freeze-dried materials.

Figure 6.4. Use of a Table to Present a Business Process

Monthly Departmental Financial Report Preparation Process

Stage	Action	Responsibility
1	Run the system monthly financials for each department.	Data steward
2	Import departmental financials into Excel template and check for missing data.	Assistant
3	Calculate monthly metrics and quarterly forecasts.	Analyst
4	Prepare summary graphs.	Analyst
5	Format spreadsheet and graph reports for distribution to senior management.	Assistant
6	Review the formatted reports.	Analyst
7	Distribute formatted reports to senior management.	Assistant

Layout of Process Content on Manual Pages

The page layouts for process tables and flow diagrams are similar. The page heading should be prominent and titled "How Xs Are

Figure 6.5. Use of a Flow Diagram and Table to Present a Technical Process

Process for Preparation of Freeze-Dried Material

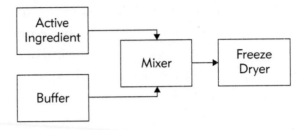

Event	Description
1	Active ingredient and buffer (inactive ingredients) are pumped into the mixer.
2	The mixer combines all ingredients and agitates the mixture.
3	The mixture is pumped into the freeze dryer.

Processed" or "How the X Works." A brief introductory paragraph might state the relevance of the process to the overall lesson. The process table or flow diagram should have an identifying marginal label such as "Work Order Process" or "Equipment Transfer Process." Figure 6.6 illustrates a financial report preparation business process laid out on the page.

Teaching Processes in e-Learning

Presenting process information on a computer is similar to presenting it in training manuals, in that the goal is to clearly depict each major stage involved. However, you have some capabilities not available in print media such as animation, sound, and zoom effects. In addition you need to lay out information to fit screen real estate rather than 8½-by-11 pages. Figure 6.7 illustrates a screen from an Internet lesson on HIV. Color and animated diagrams along with text are used to describe the process of the HIV virus life-cycle as it infects a cell. This example uses a frame layout in which

Figure 6.6. A Manual Page with Business Process Summary and Practice

The Financial Report Preparation Process

Introduction	All departments prepare monthly financial statements. These statements are formatted for readability and then sent to senior management.
Process	The table below explains how the monthly financial statements preparation process works.

Stage	Action
1	The **data steward** runs the system monthly financials for each department.
2	The **assistant** imports departmental financials into the appropriate Excel template and checks for missing data.
3	The **analyst** calculates monthly metrics and quarterly forecasts.
4	The **analyst** prepares summary graphs.
5	The **assistant** formats spreadsheet and graph reports for distribution to senior management.
6	The **analyst** reviews the formatted reports.
7	The **assistant** distributes the formatted reports to senior management.

Practice	The report results contain an error. Identify two possible sources of the problem.

Figure 6.7. Internet e-Learning Showing High-Level HIV Lifecycle

Accessed from www.roche-hiv.com January 4, 2007

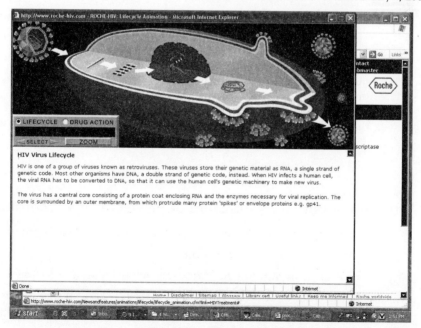

Figure 6.8. Internet e-Learning Showing Detail of the HIV Life Cycle

Accessed from www.roche-hiv.com January 4, 2007

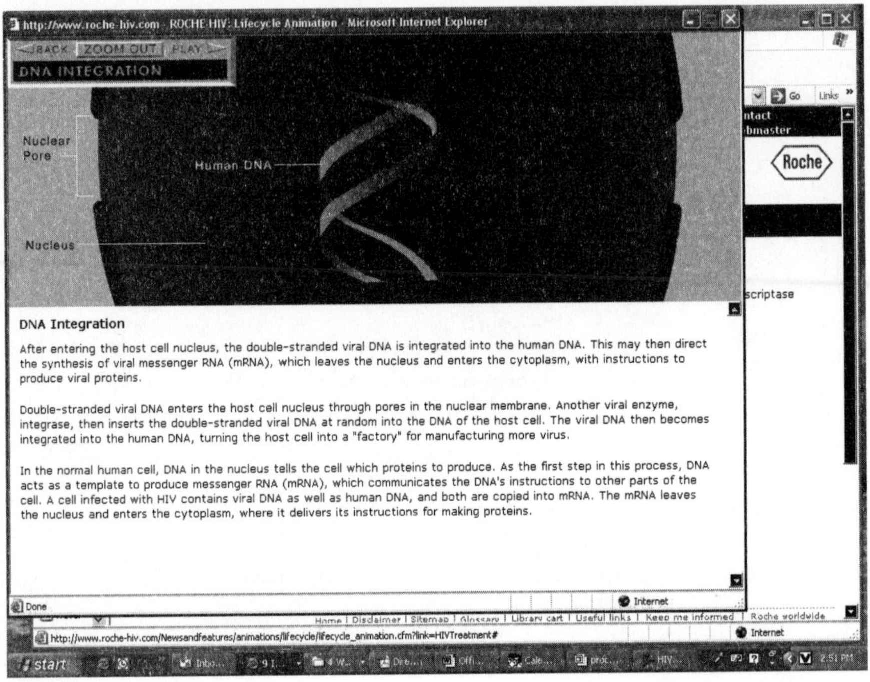

the scrollable text frame at the bottom of the screen can be viewed independently of the animated graphic at the top. The top diagram offers an overview of the entire lifecycle. The various stages are cued with the arrows. Upon clicking on an arrow and selecting the zoom button, a detailed illustration and descriptive text of that stage are provided, as illustrated in Figure 6.8. This is an especially effective technique to present layers of process details while still maintaining an overview of the bigger picture.

Using Animation and Narration of Processes in e-Learning

Animation is another feature that can be used to present processes that involve equipment or other elements that can be graphically illustrated in multimedia. For example in Figure 6.9 I show one frame from a process lesson on how toilets work. When the user

Figure 6.9. Online Animation of How a Toilet Works

Accessed from http://home.howstuffworks.com

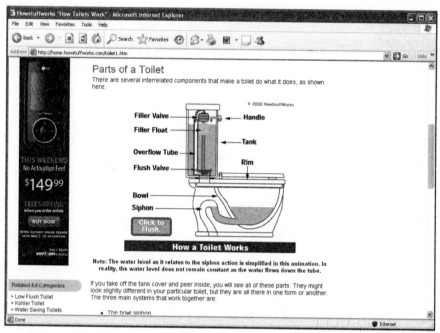

selects the "*click to flush*" button, the visual animates to illustrate the flushing cycle. This example uses text to describe the flushing cycle. Research recommends that, when possible, you describe animations with audio narration to avoid split attention between two visual sources, that is, the animated visual and the text. The part labels in this example are effective because they are positioned close to the relevant portion of the graphic.

You might think that presenting a process with animations would be advantageous compared to a static series of illustrations. However, controlled research comparisons of dynamic versus static process displays (including a lesson on how a toilet flushes) show a learning advantage for static visuals over animated displays. Static visuals may be better because (1) animations present so much information rapidly that learners experience mental overload and (2) static diagrams require the learner to engage in a mental animation process, which promotes deeper learning (Mayer, Hegary, Mayer, & Campbell, 2005). Based on evidence to date, I recommend that, whether presenting a process in a manual or in e-learning, you allow the learner to control the rate of presentation of information by using a series of static displays or, if using animation, allow learners to control presentation rate between state changes. In addition, use audio to describe stages, especially in complex visual displays such as animations.

Process Simulations in e-Learning

In Chapter 3, I discussed using simulations to teach procedures related to the use of computer software or equipment involving dials and buttons that could be displayed on a computer. Simulations can also be used to illustrate processes. In a simulation the learner interacts with a program that emulates the real process and learns about it in an interactive way. For example, a simulated steam generator might include valves that could be manipulated and gauges that would change in response to different stages in the process. The learner would "see" the process in action and experience it more directly than through a description.

Simulations are very popular with learners, but they require more time and effort to create than the more standard techniques described above. We do have evidence that procedure simulations are useful and cost-effective ways to give hands-on practice with tasks that are difficult to practice in the workplace. However we *do not* have data that simulations are a more effective or cost-effective route to *build process knowledge*. Until we have more data indicating when and how to design process simulations that teach effectively, I recommend sticking with more traditional methods such as a series of static diagrams accompanied by application-level questions.

Practice Methods for Processes

In planning practice for process content, think through your purpose in presenting the process. Sometimes you won't need to design any practice. In the lesson on how to brush your teeth, for example, the outcome of the lesson is to train learners on correct procedures for tooth brushing. The information on the process of tooth decay is intended primarily to provide greater understanding of why tooth brushing is important. The decay process is included as a "nice-to-know" topic in the lesson.

In contrast, if work performance will be substantially aided by an understanding of the process, practice is warranted. Some knowledge work, including such tasks as system design or troubleshooting either equipment or customer problems, are common examples. Knowledge of how the erylitizer works is necessary for the erylitizer maintenance engineer. Programmer coding will be directly improved by an understanding of the control and flow processes of the software.

Practice should be designed to match the learning objective at the application rather than the remember level. Don't ask the engineer to list the five major stages in the operation of the erylitizer. Instead, develop an exercise, like the one in Figure 6.10 that matches the first objective in Table 6.1. This objective and exercise requires erylitizer problem solving based on a description of malfunction. Similarly, suppose a customer calls in upset about an overdue notice on her bill. To respond effectively, the customer service representative must be able to apply her knowledge of the billing cycle to this particular customer's situation. Practice exercises presenting a variety of customer queries would help the representative apply her knowledge of the billing process.

The key to the design of process exercises for employees involved in problem solving is to collect realistic case studies of common malfunctions, customer questions, and so forth that can be converted into practice exercises. Ask your technical experts to collect these types of incidents that could be converted into case-study problems for practice exercises.

Design of e-Learning Practice

The same guidelines used for the design of process practice in the classroom apply to e-learning. However, since the computer cannot

Figure 6.10. Design Exercises That Require Learners to Apply Process Knowledge in Job Context

Solving Glopple Problems

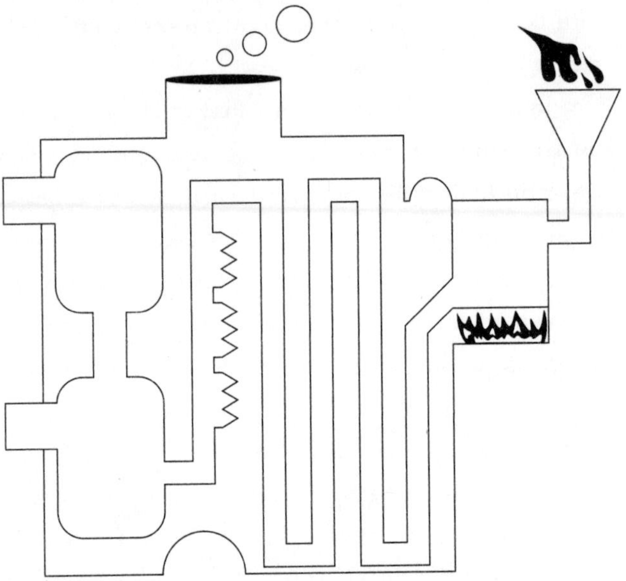

Exercise: State the probable malfunction based on each scenario and describe what diagnostic steps you would take:

1. Glopples exiting from acid bath look like this: ⸙ ⸙
2. Glopples lack tensile strenth. Heat reactor warning light is on.
3. Glopples exhibit greenish hue on surface.
4. Oil pressure valve is below normal and uncharacteristic gurgling sound is noticed by operator.

easily analyze responses to open-ended questions such as "Describe what would happen if there were leakage between the right and left ventricle," more structured formats are required. For example, in the toilet flushing process lesson, a malfunction could be described and the learner could be asked to click on the component or components that caused the problem. Or a faulty stage could be described and the learner asked to select from several options what the most likely result would be to the system.

Practice Using Simulations

If you are using an authoring system with simulation capabilities, some creative exercises could be designed capitalizing on those features. For example, suppose you asked: "What is the likely result of leakage between the right and left ventricle?" After selecting an answer, the learner's response could be "played out" in the simulation as feedback. However, as mentioned previously, we do not have evidence that simulations are more effective than more traditional approaches to teaching process knowledge. While these kinds of training systems can be very engaging, they can also be expensive to build. The knowledge and skills gained may not have much transferability beyond the specific process or equipment being simulated. As a general rule, start out with a simpler instructional approach and evaluate outcomes on job performance before investing in elaborate and expensive simulations.

Evaluating Learning of Process Knowledge

This brings us to the question of how to demonstrate that you have been successful in building process knowledge to learners. First, consider your instructional objective. If the major objective is to perform a task and the process information is provided as background information, you need not write a learning objective or provide practice or evaluate learning.

However, if the process knowledge is critical to task performance or is the major intended outcome of the lesson, evaluation is needed. Use formats similar to the practice exercises you have designed. If you have been training customer service representatives to solve customer billing problems based on their knowledge of the billing cycle, generate test items that provide customer questions to be resolved. If you have been training maintenance technicians how to repair equipment or software, generate failure scenarios that will require application of process knowledge. While the formats should be similar to the practice exercises, vary the actual scenarios involved.

How to Teach Principle-Based Tasks

We have now reviewed how to teach four of the five major types of content in all training systems. The last type I will address is tasks based on principles. Our discussion of principles will lead us to two categories of training outcomes: near transfer and far transfer. With increased emphasis on knowledge work, far-transfer tasks are of growing importance to organizational performance.

In the next chapter I will discuss the issues to consider when deciding how and when to teach principle-based tasks as part of your instructional system. To help you in situations in which you do need to train principle-based tasks, techniques to present and practice at the application level will be stressed.

For More Information

Mayer, R.E., & Clark, R.C. (2007). *e-Learning and the science of instruction* (2nd ed.). San Francisco, CA: Pfeiffer. (See Chapter 3 on the Multimedia Principle.)

Mayer, R.E., Hegary, M., Mayer, S, & Campbell, J. (2005). When static media promote active learning: Annotated illustrations versus narrated animations in multimedia instruction. *Journal of Experimental Psychology: Applied,* 11(4), 256–265.

Figure 7.I. The Content-Performance Matrix: Principles

	Facts	Concepts	Process	Procedure	Principle
Apply		Classify new examples *Select the valid signatures*	Solve problems Make inferences *Given these symptoms, what could be wrong?*	Perform the Procedure *Log on to the system*	Perform the task; Solve problems *Close the sale*
Remember	Remember the Facts *Write out the password*	Remember the definition *List the features of a valid signature*	Remember the stages *Describe how the Erylitzer works*	Remember the steps *List the steps to log on to the system*	Remember the guidelines *List the Guidelines to Close a sale*

7

How to Teach Principles

CHAPTER OVERVIEW

In Chapter 3, I described how to train procedures, the basis for many work-force tasks. Procedures consist of a series of steps that are performed more or less the same way each time the task is done. However, many job tasks are not procedural in nature. There are no exact steps that can be specified because the situation or context in which the task takes place will be different each time. I call these *principle-based* or *strategic* tasks. For example, sales activities are typically not performed effectively in a procedural manner. When making a sale, the associate needs to assess the customer to define his or her needs and then offer product proposals that meet those needs. How this is done will vary with each customer and will depend on the associate's relationship with the customer and the current product features as they apply to that customer. In short, the sales associate will need to use judgment to apply guidelines of an effective sales cycle to each customer in a unique way.

In this chapter I distinguish between routine tasks that are procedural (also called *near-transfer* tasks) and those that are principle-based or *far-transfer* tasks. Since identifying the guidelines that underlie principle-based tasks is more challenging than defining procedural steps, I will offer some suggestions to guide your task analysis. I will then summarize the key instructional methods needed to teach far-transfer tasks and to practice them at the application level. We will see how to display the key content and practice exercises for far-transfer tasks both in print and e-learning media. Last, I will summarize how to assess the learning of principle-based tasks.

Knowledge Work and Principle-Based Tasks

While job performance in the first half of the 20th century relied heavily on procedural work, knowledge work increasingly accounts for a larger proportion of organizational expertise. "The center of gravity in employment is moving fast from manual and clerical workers to knowledge workers who resist the command-and-control model that business took from the military one hundred years ago" (Drucker, 1998). A knowledge worker is primarily a specialist who directs and disciplines her own performance through organized feedback from colleagues, customers, and supervisors. Much knowledge work is grounded in tasks that are *strategic* or *principle-based*. In 2006 the highest share of workforce training funding went to domains that rely on principle-based tasks: sales, management and supervision, and IT-systems training (Industry Report, 2006). In addition, performance support systems are increasingly augmenting procedural training with embedded step-by-step working aids. Performance support in some cases can even replace the need for procedural training. In the age of the knowledge worker, much of the strategically critical work that makes organizations productive relies not on procedures but on principle-based tasks.

Near- and Far-Transfer Training

Compare the work in a fast-food restaurant with that of a professional chef. The focus in fast food is consistency, speed, and quality. The customer is expecting a similar product to that bought at another restaurant in the chain, as well as fast service. Fast-food franchises are not interested in workers using judgment or creativity to design the hamburger or sandwich; what is needed is a fast, safe, and reliable mechanism for producing a quality product. Operationally, these types of organizations rely heavily on consistent, accurate application of procedures.

In contrast, a chef's work involves knowledge of the target audience, principles of food preparation, and creative design to prepare unique "signature" cuisine. For example, in chef training, a unit on cakes will require learners to make several cakes, leaving out a key ingredient each time. This gives the student an understanding of the principles of baking—how the eggs, salt, flour, baking power, etc., influence the outcomes. In contrast, training at a fast-food restaurant focuses on application of step-by-step working aids and pre-prepared components to quickly and accurately produce a consistent food item. In short, the focus in procedural learning is on the **how to,** whereas the focus in principle-based learning is the **what, why, and how.**

As illustrated in Figure 7.1, the distinction between step-by-step training and training that involves application of guidelines to changing contexts can be summarized in the concepts of *near-* and

---

Figure 7.1. Near- and Far-Transfer Tasks

Near Transfer	**Far Transfer**
Steps performed same way each time	Guidelines adapted to the situation

far-transfer performance outcomes. Near-transfer training involves the teaching of tasks that are procedural. That is, you can show the employee the exact steps he should take to achieve the desired outcome. In near-transfer training, the skills are applied more or less the same way each time the task is performed. Logging onto the computer is an example. By contrast, far-transfer tasks are performed under circumstances that are dynamic. Because the actual steps must be adapted to fit each situation and will be different every time, a more robust type of learning must take place—one that enables workers to adapt guidelines to different work-related contexts.

Distinguishing Between Near- and Far-Transfer Training

Most jobs involve a combination of near- and far-transfer tasks, but one of the two types usually predominates or is more critical to the organization's strategic goals. For example, while the fast-food franchise worker may use step-by-step procedures to prepare a food item or ring up a sale, he or she may also need to apply principle-based skills as part of the customer service duties or in conjunction with store management. Because the instructional techniques differ, you need to decide which of your training goals call for near-transfer and which for far-transfer learning. Ask yourself: Does the instructional goal require your trainees to follow a prescribed set of steps in a relatively consistent environment? Or will the task circumstances be changing sufficiently each time that you need to train learners to effectively apply guidelines?

In most organizations far-transfer performance deals with one of the following domains: (1) application of social skills, for example, sales, client service, and management; (2) design of new products or services, such as product design or marketing; and (3) analysis, such as financial planning, or non-routine problem resolution, for example, advanced troubleshooting or diagnosis.

There are advantages and disadvantages to both types of training. Near-transfer tasks are easier to train and give you a good chance of getting the specified result if you follow the appropriate teaching steps. On the other hand, while the employee will be well equipped to perform the tasks as they are trained, if task-related circumstances change, their ability to adapt will be limited.

By contrast, far-transfer tasks are more challenging to instruct. Because the employee will have to adapt the guidelines to fit the various situations in which they will be applied, the likelihood of success is less than with near-transfer tasks. On the other hand, once the skills are acquired, the employee is more versatile because she can use judgment to make adjustments as circumstances dictate.

Sometimes you may have to decide whether to train at a near- or far-transfer level, depending on the level of the employee. Teaching electronic troubleshooting is one example. You may want entry-level employees to quickly learn to perform routine troubleshooting procedures that will resolve 75 percent of the common problems. They will know what signs to look for and what components to replace based on those signs, with little understanding of why. As they won't need much knowledge of electrical theory, their training will be designed around flow-chart types of decision procedures. In contrast, the organization may also need more advanced professionals who can form hypotheses and resolve problems not encountered before. This requires a greater depth of knowledge to support the application of principles to various unpredictable situations. Their training will be principle-based and will incorporate substantial background knowledge such as device components and processes, basic concepts such as electrical theory, and troubleshooting heuristics.

Check Your Understanding

To be sure you can distinguish between near- and far-transfer tasks, try the short exercise in the Appendix under Exercise for Chapter 7.

Far-Transfer Training and Guidelines

If you are going to be training far-transfer tasks, you will be working with principles that underlie or make up the guidelines to be trained. A principle is a cause-and-effect relationship that results in a predictable outcome. The most valid principles are based on scientific research from the physical, biological, or social sciences or from abstractions based on best practices. For example, "Specific challenging assignments accompanied by informative feedback and appropriate consequences lead to greater workforce productivity" is a cause-and-effect statement that has been verified by research (Clark & Nguyen, in press). This principle can form the basis of guidelines for management training sessions in which supervisors and managers learn to construct specific and challenging assignments for their staffs and to provide regular and informative feedback and consequences. However, what constitutes a specific and challenging assignment varies from task to task and from worker to worker. Such guidelines cannot be taught as procedures. The desired outcome of your training will be to help workers convert guidelines into situation-specific procedures in the workplace, as illustrated in Figure 7.2.

Figure 7.2. Translating Principles into Context-Specific Procedures
General Guidelines

Make task assignments that are specific and challenging

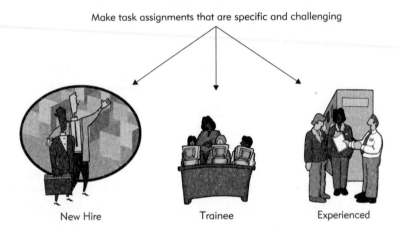

New Hire   Trainee   Experienced

Identifying Valid Guidelines

Perhaps one of the consistent gaps I see in workforce learning is failure to validate the *guidelines* that make up far-transfer tasks. Spending the time and effort to train far-transfer skills such as management and sales based on guidelines that **are not valid** will **not** translate into operational advantages. To qualify as valid, the guidelines, when applied, should lead to job actions that accomplish desired bottom-line results. For example, your goal might be to prepare sales staff to write an effective proposal or research analysts to identify and organize relevant information for a policy report. Unless you base your guidelines on the activities validated to result in effective proposals or reports, it is only by chance alone that you will reach your goals. How do you identify the guidelines that will make up the basis for the far-transfer tasks you train?

Generally, it is not too difficult to identify the steps good performers typically follow to complete a procedural task. But you cannot do this as easily for far-transfer tasks for two reasons. First, there is rarely one right approach to far-transfer tasks—and therefore no single expert can demonstrate the guidelines. Second, many expert performers cannot readily articulate the way they tackle far-transfer tasks or the reasons for their actions. Experts will typically state that their work is "intuitive." In far-transfer training, you will have to do additional research to identify the appropriate guidelines to be trained. I will briefly describe some techniques here.

Figure 7.3 summarizes two general approaches you can use to define the guidelines you need: (1) research of external documents and experts and (2) abstracting best practices from work products or top performers either via internal or external benchmarking. A combination of both techniques will give you the best data from which to derive the guidelines most applicable to your unique context.

Figure 7.3. Defining Critical Far-Transfer Task Guidelines

Document Search and Analysis Expert Interviews	Analysis of Best Practices

Drawing on External Research

In many domains you can find scientific research that provides guidelines applicable to various business problems. For example, from the social sciences, studies on how to maximize employee productivity reveal the following guidelines:

- Provide specific and challenging work assignments.
- Collect and communicate informative feedback on the quantity and quality of work outcomes.
- Provide positive consequences for desired actions or outcomes.

In environments in which work assignments are vague or undemanding, where the results of work activities are never known, or where consequences or incentives are unrelated to activities or results, productivity will decline. These guidelines could provide you with the basis for part of a management/supervisory training program. If you are not an expert in the area to be trained, you might locate a recognized expert who could help you identify the most relevant research guidelines. Contact your local university or professional society for resources to locate individuals with expertise in a particular field. If you are a technical expert, you are likely already familiar with the key principles of your field. You may need only to identify those that are most applicable to your desired training outcomes.

A second approach is to abstract the basic principles or guidelines by analyzing the actions or work products that reflect best practices either within your organization or external to it via benchmarking. I summarize three methods for deriving guidelines based on best practices including:

1. Performing structured analysis of best practice work products or activities;

2. Analyzing stories of best practices/products from master performers; and

3. Performing detailed analysis of the decisions experts make when solving problems.

Identifying Guidelines from Best Practices

Your first step will be to identify employees, teams, or products that embody best practices and compare them with the work of those

whose practices are new or average. Try for an independent validation of workers or teams that reflect best practices. Independent validation means you have an objective measure of performance—not
just someone's opinion. Sales volume might be a good indicator of
effective performance. Instructor rating sheets would provide good
validation of instructors who are perceived as effective by their students. Student achievement results would be a valid indicator of
instructional effectiveness.

It is critical to observe several master performers in order not to
misconstrue one top performer's idiosyncratic techniques as an important feature. You need to look at commonalties among the top performers **not shared** by new or less-proficient workers. For example,
an evaluation of several highly rated instructors showed some to be
dynamic and energetic, while others were quieter in their classroom
presence. On the surface their instructional practices seemed quite different. What turned out to be a critical underlying factor of instructor
success was the individual's ability to sustain attention in a supportive
manner. Some instructors did this with humor, some with animated
presentations, and others with interrogative teaching techniques that
kept the participants actively involved. Below I summarize three methods you can use to derive guidelines from validated best practices.

**Structured Observations: Deriving Guidelines from Looking and
Listening.**   Much can be learned by observing "power users" or best
practitioners, with the observation supplemented by an interview.
This is practical when the performance of critical tasks is overt
(for example, can be seen and/or heard) and when work is relatively
short term (that is, you can observe multiple examples in a matter
of hours). Ideally two or more observers watch work in progress,
such as customer service representatives dealing with customer
issues. Often a taxonomy of tasks will evolve or may have been
previously established. For example, you may identify five basic
client interaction types. Guidelines will need to be derived for each
of these task types. For example, guidelines for effectively handling
a routine tax payment question might vary drastically from those for
handling an irate taxpayer who has received an audit notice.

In some work situations, you can infer guidelines by analyzing the products of the work. For example, if the main work output is a report, you can review a number of highly rated reports and derive the common features among them that made them successful. When you can access a tangible work product, start by evaluating those products, since this type of analysis is least obtrusive to the work in progress and can be completed asynchronously. In other situations, however, you will need to observe the work of the top performers or top-performing teams in action. For example, in sales and customer service, there may not be any tangible product that will offer insight into the relevant guidelines.

The Critical Incident Technique: Deriving Guidelines from Stories.   Sometimes it is impractical to observe work in progress due to confidentiality of the setting, length of time to complete tasks, or the fact that much of the work is mental so there is little to be seen. A relatively quick technique to define guidelines in these situations is to gather stories. Ask best performers to identify materials that formed the basis of a recent successful case they worked that might serve as an exemplar for others. For example, ask them to bring all relevant records such as computer files, customer descriptions, interim notes, and any final output. Ask respondents to capsulate these cases into twenty-to-thirty-minute stories. Video-record the stories. Review the recordings—ideally with a team that includes subject-matter experts and instructional experts as well. Look for commonalties in inputs to the case, responses to those inputs, and actions or products that resulted. For example, best-practice auditors may bring one or two cases they considered especially effective. By reviewing the data they provide, including the computer activities they undertook, the worksheets they completed, the interviews they conducted, and the final decisions/actions they generated, you can build the guidelines of effective auditing. Be sure to save the stories and the artifacts, as these can be the basis for examples and exercises that you design for the training.

The Worked-Problem Technique: Getting Inside the Mind.   The critical incident technique likely will provide you with a broad range of instances from which you can abstract high-level guidelines. If you need to drill down to greater detail, especially into how an expert thinks during task performance, you may want to try the worked-problem technique. To do this you need to have a problem or set of problem materials that are typical of the work environment and that require the application of guidelines that are relevant to your learning goals. You can derive problem cases from the critical incident meetings described above. Ask a series of experts (individually) and junior workers to solve the test problem using normal work references and tools. As they are working through the problem, ask them to talk aloud. Video-record the actions and verbal explanations. Then compare the solutions and solution processes of expert and junior workers to derive the thought processes that underlie effective problem resolution. This type of study is time-consuming. I recommend it only if definitions of detailed thinking patterns are needed that are directly linked to significant mental tasks that cannot readily be obtained through other means.

Lajoie and her colleagues (1998) conducted this type of analysis in order to build a medical training simulation that focused on surgical intensive care nursing. They started their analysis with interviews of three head nurses from the intensive care unit to identify the most difficult aspects of their work. From these interviews, job competencies that distinguish expert from beginning practitioners were defined. A second round of interviews with expert nurses identified specific case problems that would reflect the key competencies identified in round one. Once cases were written, the actions that experienced nurses would take were defined by asking three nurses unfamiliar with the case to talk aloud as they solved the case problems. As the nurses described an action they would take, the interviewer asked the reason for the action. Then, based on known responses to those actions, the interviewer would state the patient response to the action and ask the nurse to interpret that response. A sample from one of these interviews is displayed in Figure 7.4. The transcripts collected from these problem-solving

Figure 7.4. Interview to Get Inside the Mind of an Expert Nurse

From LaJoie, Azevedo, and Fleiszer, 1998

Interviewer	:	Which body system would you start with?
Expert	:	Neurology system
Interviewer	:	Why?
Expert	:	I want to see if the patient is conscious.
Interviewer	:	What would you do first?
Expert	:	I would use my flashlight to examine reaction of the pupils
Interviewer	:	The pupils both react equally to the light stimulus by contracting. What does this result imply or mean? How do you interpret this?
Expert	:	There is no brain damage….

sessions were grouped into categories, including hypothesis generation, planning of medical interventions, actions performed, results of evidence gathering, and interpretation of results.

Combine Research Approaches

In general, it's a good idea to use a combination of techniques, including literature reviews or interviews of recognized experts and observations, with interviews of independently validated top performers and/or analysis of top work products. A combination of techniques will help you see how well the research data fit the environment of the group to be trained. As a secondary benefit, the direct involvement of the target group will promote greater buy-in to the resulting training.

Learning Principle-Based Tasks at the Remember and Application Levels

As we have seen in prior chapters, all the content types except facts can be taught and learned at the remember and the application levels. I have emphasized the importance of the application level throughout. Principles are no different. At the remember level, the worker can recall or recognize the guidelines. For example, the sales trainee can state: *"First I need to form a relationship with the customers. Then I need to assess their current business needs. After that I can begin to introduce products that address their needs and fall into their range of price sensitivity."* This response tells you that the trainee has a good memory for the guidelines. It does not mean she will be able to implement them in a sales encounter. At the application level, the learner must perform tasks in ways that are congruent with the guidelines presented. Because each customer scenario will differ, she must learn to adapt the successful guidelines to the differences in each case. Thus at the application level, the learner responds to realistic case scenarios for which she would adapt the guidelines listed above.

Writing Far-Transfer Learning Objectives at the Application Level

If your lesson tasks are far transfer, your main lesson objective should be written at the "application-of-a-principle" level. Instead of a remember-level objective such as "List the five guidelines for an effective sales engagement," an objective at the application level would read: "You will be given a specific product, product catalog, and customer description. You will role play initiating the engagement, using the guidelines for initiating the sales process." If you were designing training to teach technical instructors effective virtual classroom skills, your objective might read: "You will present a twenty-minute lesson in the virtual classroom. Your lesson must engage learners through explanatory visuals and frequent application level interactions that use at least four of the seven response facilities." Notice that principle-based learning objectives ask the trainee to respond to realistic job scenarios by appropriately applying the guidelines.

Supporting Learning Objectives

In addition to your main lesson objective that requires the learner to apply guidelines while performing the job task, some supporting objectives are often necessary in far-transfer training. In our sales training course, it would be helpful to ask learners to identify appropriate and inappropriate applications of the guidelines before trying them out themselves. Thus, a supporting objective might state: "You will watch ten videotaped sales engagements. For each you will decide whether the guidelines were applied effectively and state what improvements, if any, are needed." The virtual classroom facilitation skills lesson might focus on identification of different synchronous e-learning response facilities with an objective that asks learners to classify a series of sample instructor-student exchanges.

In addition, if your far-transfer task relates to problem solving involving equipment or business systems, you are likely to include a process-supporting objective. For example, for troubleshooting, you would include objectives that focus on equipment operations. For sales you might include objectives that focus on the sales cycle.

Once you have identified a valid set of guidelines and written your learning objectives, your next challenge will be to develop training that enables employees to apply them to the diverse changing situations they will face.

Training Far-Transfer Tasks: Overview

Recall that the strategy for teaching procedures is to provide the steps and a demonstration accompanied by step-by-step practice until the employee performs the task confidently. Because the application environment is pretty similar each time, procedures can be taught with a demonstration that emulates the work environment.

However, in far-transfer training the approach must be different. Your goal is to build mental models that can be adapted to changing situations. To accomplish this, first present the guidelines, followed by several varied context worked examples that illustrate how the guidelines can be applied in diverse circumstances. Varied context worked examples show how a far-transfer task performance might play out in the alternative situations the employee is likely to encounter. To ensure that learners process these examples, add some self-explanation questions. Questions can require learners to identify the guidelines or principles applied in the various stages of the example or can ask learners to compare similarities in two examples. In addition, some non-examples of applying guidelines are often useful. Learners can observe sample task performances and identify which ones effectively implement the guidelines. Discussion of ways to improve the ineffective examples should be designed into the exercise. I also recommend analogies when you can derive a metaphor that will help learners quickly grasp the new knowledge. Finally, the trainee must practice applying the guidelines in a series of exercises, usually presented in the form of case studies. The next sections will take a more detailed look at how to train principles in classroom and in e-learning environments. As with all five content types, training techniques are of two major types: information displays and practice exercises.

Training Far-Transfer Tasks

Regardless of the delivery media (classroom or computer), you will use similar instructional methods that include statements of the guidelines, worked examples that elicit explanations or comparisons, non-examples, and sometimes analogies. In addition you will assign practice that requires learners to apply guidelines to various job-realistic case scenarios. You can take two approaches to the sequencing of these methods: *instructive* or *inductive*.

An Instructive Approach: State the Principle and Guidelines

Begin with a statement of the principle itself, followed by the guidelines that flow from it. You may also want to include an elaboration of the rationale for the principle. If you were teaching supervisors how to deal with an employee with personal problems, the general principle is: "You want to maintain a constructive relationship with the employee, and you are not legally qualified to give professional advice on personal problems." Guidelines that follow include:

1. After the employee has presented the problem, respond with paraphrasing and empathy so the employee knows you heard her.

2. Tell the employee you are not qualified to give advice on these matters.

3. Refer the employee to the appropriate company or community resource for professional help.

4. Follow up by asking whether help was obtained and how things are going.

An Inductive Approach: Derive the Principle and Guidelines from Examples

Rather than stating the guidelines, in an inductive approach, you start with some examples that show the guidelines being implemented in various settings. For example, three or four video vignettes of workers in different settings show the employee asking the supervisor for advice on dealing with a spouse problem, a medical problem, or an addiction problem. As they view each sample, learners take notes and then compare and contrast the three samples by defining the underlying guidelines. In this approach learners are actively engaged in deriving the guidelines based on the examples they view. The inductive approach is more learner centered than an instructive lesson, but also generally requires lengthier training time.

Develop Varied Context Worked Examples

Whether using an instructive or inductive sequence, use a series of examples that incorporate diverse settings to show the guidelines being implemented. The supervisory training program might include several videotaped examples. Other examples can be presented in manuals or on screens using words and visuals. It will be important to get learners actively engaged in the examples, and I will show you how to do so in the sections to follow.

Develop Varied Context Non-Examples

After showing a variety of situations in which the guidelines are effectively applied, you might present some non-examples. It will be important to build in cues that show how the non-examples ignore the guidelines. In our supervisory training program, a video non-example might show a supervisor describing his experiences in his own divorce or relaying how his best friend resolved an alcohol problem. The instruction would need to show that, while the supervisor is attempting to be supportive, he or she is not qualified to give this type of assistance and could be held legally liable, and that the employee is entitled to the most competent professional resources available. For these reasons, the actions taken violate the guidelines.

Analogies: A Key Information Display

After a statement of the principle and guidelines accompanied by varied context examples and non-examples, an analogy can be a powerful technique to help learners build new mental models. Analogies are efficient because they call on existing knowledge to explain a new relationship. For adding fractions, the slice-of-pie analogy is commonly used to show how two-fourths add up to one-half. For teaching fundamentals of voltage and resistance, an analogy of water moved by various-sized pumps through pipes of different diameters is often used.

Many introductory computer courses have drawn on the analogy between typical office concepts, such as file folders, in-and-out baskets, and filing cabinets, and computer concepts, including records, files, and temporary and permanent memory storage.

While analogies are a very effective way to teach principles, coming up with the right analogy is often very difficult. First, the basis for the analogy must be known already to the trainee. Second, the critical features of the analogy must map functionally onto the new guidelines to be taught. If you are presenting a complicated analogy, it will be important to point out how the elements of the analogy map onto the new knowledge.

Teaching Far-Transfer Tasks in the Classroom

We have seen that in the teaching of principles a statement of the principles and guidelines accompanied by diverse worked examples and non-examples make up the major informational displays. In addition, use analogies when practical. Figure 7.5 is drawn from a lesson on how to prepare effective and legal hiring interview questions. Note the topic starts with a statement of the principle and a bulleted list of guidelines. The first example provides a short hiring scenario and walks the learner through an example in which first technical skills and then performance skills were identified, followed by construction of the questions. The next page of this topic includes samples that focus on a different context from the first example.

Figure 7.5. Sample Principle Topic from Lesson on Planning the Hiring Interview

How to Write Hiring Interview Questions

Principle	To develop fair and effective interview questions, you must focus on the candidate's bona fide occupational qualifications.

Guidelines	Follow the guidelines below when developing interview questions.
	• Determine the required technical and performance skills that you will investigate during the interview.
	• Formulate a question for each selected technical and performance skill.
	• Write questions that are
	   – job relevant,
	   – open-ended,
	   – behavioral-based, *and*
	   – non-discriminatory.
	• Review questions to ensure they meet the above guidelines.

Example 1	Lynn wants to fill a vacant executive secretary position in your office. She completes the job analysis, lists the required technical and performance skills, and highlights those skills she plans to investigate during the interview.

Technical Skills	Performance Skills
Operate SUN terminal	Deal with senior manager
Operate STU-III and fax	Manage office staff
Use FrameMaker software	Follow through on tasks and report on results
Operate phone system	Cope with simultaneous inquiries

Here are sample interview questions Lynn prepared for the interview.

- In our office, we work exclusively with FrameMaker on SUN terminals. Describe a situation in which you had to format a large document in FrameMaker on a SUN, and tell me how you did it.
- This position will require you to deal with senior managers who may not be happy about having to wait to see me. Tell me about a time when you had to deal with someone senior to yourself who was unhappy with your message. How did you handle that?

Example 2   Mark is seeking a new instructional designer for the training organization. He completes the job analysis, lists the required technical and performance skills, and highlights those skills he plans to focus on during the interview

Technical Skills	Performance Skills
Use M/S Office software	Conduct JTA
Work with Adobe authoring tools	Lay out storyboards and manuals
Use FrameMaker software	Work as part of a team
Use synchronous e-learning tools	Communicate effectively

Here are sample interview questions Lynn prepared for the interview.

- In our department, designers create materials for both workbooks and e-learning. Tell me about previous products you have created for print and for e-learning delivery.
- This position will require you to work with others, including graphic artists, SMEs, and programmers. There are always tradeoffs between efficiency and quality. Tell me about a time when you had to make a tradeoff between efficiency and quality. How did you handle the situation?

Figure 7.6 is adapted from an experimental study that compared several approaches to teaching negotiation skills—a far-transfer task. The researchers tested learning a specific negotiation technique (a contingent negotiation strategy called a "safeguard") from several different lesson versions. The sample in Figure 7.6 gave the best results because two varied context examples were provided in words and in a diagrammatic summary and because the learner was

---

Figure 7.6. Experimental Lesson on Negotiation Strategies

Adapted from Gentner, Loewenstein, and Thompson, 2003

How to Negotiate a Contingent Contract

Instructions	In a moment, you will read two examples of an effective negotiation approach. Pay close attention as you will be asked questions about these afterwards.
Principle	A safeguard is a type of negotiated agreement in which the future is uncertain, but people are willing to proceed based on what they think will occur. It gives each party a "no risk" guarantee in a risky situation.
Example 1	An American company has ordered parts from a Chinese company:

- Normally the Chinese company sends parts by boat.
- The American company was worried that the parts wouldn't come in time for them to meet a construction deadline, so they asked the Chinese company to ship the parts by airmail, rather than sending them by boat.
- The Chinese company refused, not wanting to pay the expense.

Delivery of Parts

Chinese Co. thinks:

The boat will come on time

American Co. thinks:

The boat will come late

Compromise Solution	The two companies considered sending half the parts by boat, and half by airmail.
	• They realized however, that this was a poor solution because it satisfied neither company's needs – the Chinese company would have to pay more to send the parts, and the American company wouldn't have everything it needed.
Safeguard Solution	• Instead, the two companies decided to make a safeguard. • The Chinese company will ship the parts by airmail. Both companies will watch the boat to see when it arrives in the states. • If the boat arrives early, the American company will pay for the added cost of airmail. • If the boat arrives late, then the Chinese company will pay for the added cost of airmail.

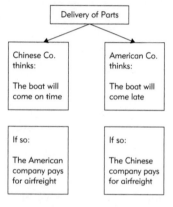

Example 2	Mark and Paul are arguing about where to stay on their spring break trip to Cancun. They are going at the peak travel time so they know that figuring out where to stay in advance is important. Mark's parents own a condo there where they could stay, but they could also reserve a hotel room. The condo would be an idea place to stay, but Mark's parents might be staying in the condo at the same time. Neither wants to spend his vacation sleeping on the floor if Mark's parents do end up coming. Mark says in all likelihood his parents won't come. Paul wants to make a reservation at a hotel just in case.
	They argue about reserving a room at a nice hotel, and leave it open as to who will pay for it: if they don't need the room, Paul will pay because he wanted to reserve it in the first place, but if they do need the room, Mark will pay the deposit because he didn't think it would be necessary.

(Continued)

(Continued)

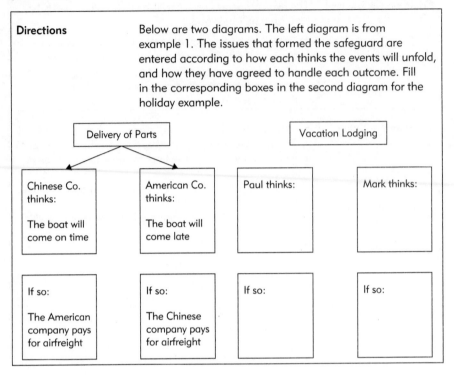

Directions   Below are two diagrams. The left diagram is from
example 1. The issues that formed the safeguard are
entered according to how each thinks the events will unfold,
and how they have agreed to handle each outcome. Fill
in the corresponding boxes in the second diagram for the
holiday example.

Delivery of Parts			Vacation Lodging
Chinese Co. thinks: The boat will come on time	American Co. thinks: The boat will come late	Paul thinks:	Mark thinks:
If so: The American company pays for airfreight	If so: The Chinese company pays for airfreight	If so:	If so:

required to make an explicit comparison between the two examples
by focusing on the similarities.

The first example focused on a safeguard solution between an
American and a Chinese company over shipping of parts. The
second example involves a negotiation over reserving a vacation
hotel room. A follow-on activity places a diagram summary of the
shipping scenario next to blank diagrams of the vacation lodging
scenario and asks the learner to complete those diagrams. Gentner,
Loewenstein, and Thompson (2003) concluded: "We suggest that
one aim for instruction should be not simply to provide cases but
to facilitate active case comparison. This may be done directly
(by juxtaposing two cases and asking for their commonalities as we
have done)" (p. 404).

Teaching Far-Transfer Tasks in e-Learning

Of the five content types, the most challenging to migrate to e-learning is far-transfer tasks. This is not so much due to the informational methods needed as to challenges in design of practice that elicits flexible responses from learners and can provide meaningful feedback. You can take two basic approaches to e-learning designed for far transfer. One is to adopt an instructive design similar to that we saw in the classroom examples described in the previous paragraphs. The other is to exploit the simulation features of e-learning and adopt a problem-based learning design that uses a more inductive immersive approach. We will take a brief look at both alternatives.

Instructive Methods for e-Learning

In lessons that follow an instructive approach, present the guidelines followed by a series of interactive examples that gradually transform into practice assignments. I diagram this approach, known as *faded worked examples,* in Figure 7.7. Note that your first example is worked completely in the lesson. Then the examples that follow are called *completion examples.* Each uses a different context and each assigns more steps to the learner to complete than the preceding example. For example, in a lesson on how to construct a relational

Figure 7.7. Fading from a Full Worked Example to a Practice Problem
From Clark, Nguyen, and Sweller, 2006

⬤ = Worked in Lesson

◯ = Worked by the Learner

Worked Example	Completion Example 1	Completion Example 2	Assigned Problem
Step 1 Step 2 Step 3	Step 1 Step 2 Step 3	Step 1 Step 2 Step 3	Step 1 Step 2 Step 3

database, our first worked example uses a video store context and asks learners to respond to questions that correspond to the data base guidelines shown in the example. Figure 7.8 shows Step 3 in our example. The learner views the step worked in the instruction and responds to the question about the step in the upper-right box. The purpose of the question is to ensure that the learner studies the example and processes it in a deep manner. A second example uses a different context—the design of a library database. In this example, shown in Figure 7.9, the instruction demonstrates part of the step and the learner is asked to complete it. A third example (not shown) uses a different context and asks the learners to construct the database on their own.

Figure 7.8. A Worked Example with a Self-Explanation Question
From Clark and Mayer, 2007

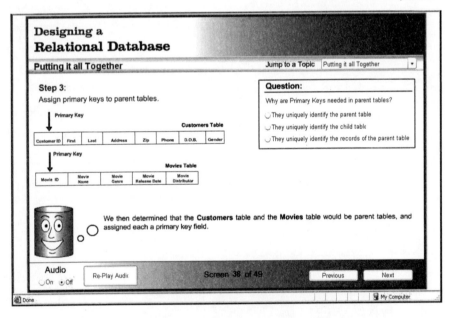

To recap, in the database lesson the learners first view a fully worked example with inserted questions to ensure review of the example. Next the learners complete part of a second example, and finally they are assigned a third different example as a full practice to

Figure 7.9. A Completion Example

From Clark and Mayer, 2007

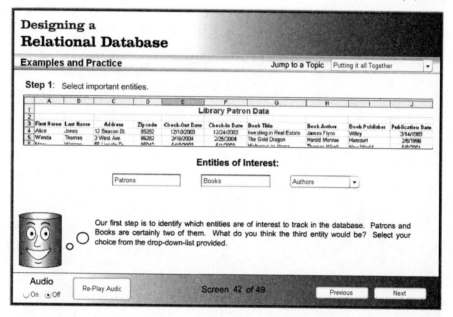

work on their own. You can read more about worked examples in resources listed at the end of the chapter.

Inductive Methods for e-Learning

An alternative approach to e-learning for far-transfer tasks is to use a simulation that will require the learner to apply the guidelines to a realistic case study. For example, in Figure 7.10 the learner has access to a number of common office tools (computer, fax, telephone, file cabinet, book shelf) to analyze a commercial loan applicant. The case begins with a video assignment from the bank manager to analyze a new loan applicant. The learner can click on any of the on-screen objects to collect data about the applicant. Ultimately, the learner completes a loan recommendation worksheet and submits it to the virtual loan committee. The learner can work in an unguided manner or can receive help from an on-screen agent.

Figure 7.10. On-Screen Resources Learners Use to Solve a Bank Loan Case Problem

With permission from Moody's Investment Services

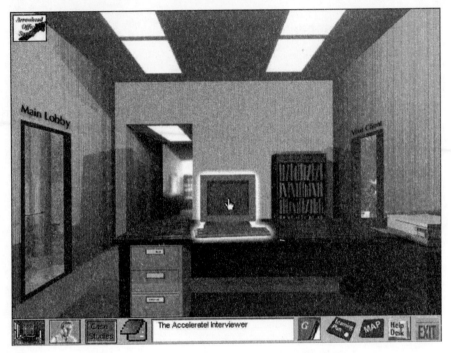

Figure 7.11 shows a screen capture from an online simulation environment developed by Lajoie to teach problem solving in a medical domain. The lesson starts with a text description of a sick patient. The learner can select words that represent symptoms they consider relevant and pull them into the left-hand evidence table. The learner can also order various medical tests, as shown in Figure 7.11. As the list of evidence accumulates, the learners can select hypotheses in the upper-left corner and rate their confidence in their hypotheses. The learning environment offers various resources, including the patient's chart, a library, and a medical consultant to help learners solve the case. At the end, the learners prioritize their evidence and compare their lists to that of a physician.

Figure 7.11. The Learner Orders Diagnostic Tests in Bioworld

Accessed October 25, 2006, from www.education.mcgill.ca/cognitionlab/
en/research/narrator.html. Reprinted with permission of S. Lajoie

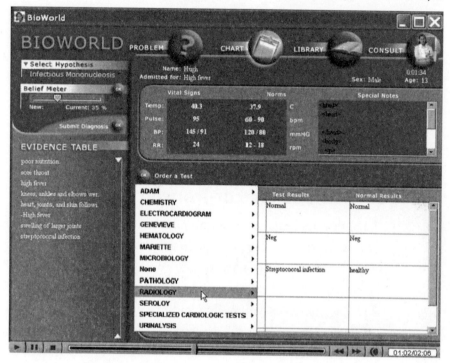

Some common features of inductive online environments designed to support acquisition of far-transfer skills are

1. Use of case problems and diverse examples for learning;

2. The opportunity to seek advice from experts while solving a problem; experts may be programmed into the lesson or may be accessed externally via email;

3. A simulation that may compress time, resulting in acceleration of learning by solving multiple problems in a much shorter time frame than the real world would allow;

4. Feedback that plays out consequences and stresses tradeoffs rather than presenting right or wrong alternatives;

5. The opportunity to reflect and try again; and

6. An emphasis not only on completing the task but on the thought and decision processes that make up effective task completion.

These instructional features differ from more instructive approaches. In fact, the approach is so distinct that I characterize it as a unique instructional architecture called *Guided Discovery*. Much has been written about guided discovery instruction, and you can read more about it in the references presented at the end of the chapter.

Practice Methods for Far-Transfer Learning

In both classroom and e-learning, practice assignments should require learners to apply guidelines to realistic work scenarios. The challenge in both settings is to assess the quality of learners' responses and give helpful feedback. Remember that there will never be a single "correct" approach to most far-transfer tasks. Therefore either the instructor or the program must find ways to determine that a student response falls within the boundaries of the guidelines and to identify the strengths and shortfalls of each response.

Design of Classroom Practice

For the lesson on how to write effective hiring interview questions, Figure 7.12 illustrates one practice assignment. The learner is provided with a hiring scenario and asked to perform the task in two stages: (1) to determine the technical requirements of the job and (2) to write interview questions that meet the guidelines. In a classroom setting, it is often useful to work on far-transfer practice assignments with a colleague or small team. Working collaboratively offers the opportunity to exchange the diverse perspectives that are characteristic of far-transfer work. After time to work on the assignment, each team could post their questions, allowing everyone to review them and to offer constructive feedback.

Figure 7.12. Practice Exercise for Lesson on Planning the Hiring Interview

How to Write Hiring Interview Questions: Practice

Practice 1 – Identify Skills

In the following scenario, identify three technical and three performance skills that you want to investigate further, and write interview questions that adhere to the guidelines on the previous page.

Job Description for Senior Intelligence Analyst:

This is a senior-level position for a certified intelligence analyst. The employee will review all incoming data, prioritize it, and assign research tasks to five junior analysts according to their ability. To do this, the analyst must be thoroughly proficient in the use of DATA and PREP in addition to available research databases. The employee will act as a coach and mentor to the junior analysts and will be expected to provide periodic performance feedback to them. Additionally, the employee will act as a liaison with the DATA and PREP systems analysts.

Technical Skills	Performance Skills

Practice 2 – Write Interview Questions

Using the skills you identified above, write at least three interview questions for any of these skills. Remember to review each question to ensure that it adheres to the guidelines.

Interview Questions	Reviewed (Y/N?)

In some settings, a more real-world type of feedback can be obtained by "trying out" the guidelines in a controlled environment. For example, in a class for new pharmaceutical sales associates, teams developed an engagement strategy for presenting the features and benefits of a new drug to a doctor. The team role played their approach to real doctors who served as consultants to the company. These doctors participated in the role play in a realistic manner and then responded with their reactions. In this way, learners received realistic feedback in a safe environment.

Project Assignments to Promote Transfer

After responding to classroom case scenarios, learners should apply their new skills immediately to job situations. For example, in the hiring interviews lesson, participants should construct and conduct actual hiring interviews shortly after the class. Following the first part of the training, participants should use actual job requirements to prepare a set of interview questions for an upcoming interview. The questions can be distributed to other class participants who can offer feedback in the classroom or later via e-mail. The value of the classroom for the practice of principle-based work is the ability to provide feedback to responses that are never clearly right or wrong, to generate alternatives, and to foster collaborative working environments that can extend beyond the training setting.

Designing Computer Simulations for Practice of Principles

Previously, I summarized computer simulation approaches to support learning of procedures. Simulations also can be used in training of principles, as illustrated in the Bank Loan and Bioworld courses mentioned previously. A simulation is a scaled-down enactment of reality in which the components will respond in accordance with the principles. Simulating scientific and mathematical principles works well. For example, in a chemistry simulation the trainee can "add" via key-presses "drops" of acid into a chemical solution graphically displayed on the screen. As the acid is added, a thermometer, a pH indicator, and the solution's color dynamically change on the screen. The learner can conduct a number of experiments without going near the laboratory.

Simulations are useful for practice-applying principles in situations that are dangerous, rare, expensive, time-consuming, or otherwise unfeasible in the normal training setting. They also allow learners to accelerate their skills by confronting or solving a number of problem scenarios that would rarely occur in the workplace and/ or that would take considerable time to complete. For example, Gott and Lesgold (2000) found that working for twenty-five hours on a troubleshooting simulator helped junior technicians achieve the same levels of competencies as senior technicians with ten years of experience. Results such as these can cost-justify the time and resources required to build effective online simulation environments.

While simulations are highly motivating and effective, they are usually time-consuming and complex to construct. They work best when there is a clear, rule-driven set of relationships so that various states in the microworld can be reliably predicted. You will need to consider the benefits derived from a computer simulation against the time and effort involved in creating it. Also keep in mind that far-transfer learning may not require a highly elaborate simulation. You may be able to achieve reasonably good outcomes with a simple simulation built using branching logic. For example, a course on medical ethics for doctors begins with a presentation of a case. The learner has links to various sources of information regarding the case and to various online experts, including lawyers, ethics experts, colleagues, and documents from the American Medical Association. In the end, the learner selects one of several links that represent a case decision and receives feedback that summarizes the tradeoffs of the decision. In this program, a few simple graphics and links to various repositories of information offer a rich yet technologically simple learning environment.

Blended Solutions: The Best of Both Worlds

Effective training delivery today will use the best mix of media to deliver those methods most suited to the instructional goal. Thus for principle-based instruction, much of the informational portion of the learning could be effectively and inexpensively handled online in asynchronous e-learning lessons showing examples of guidelines being applied effectively and ineffectively and even asking learners to respond to simplified versions of real-world problems. However, more flexible or job-realistic practice and discussion would be continued in a virtual or face-to-face classroom setting where the instructor and learner exchanges enrich the learning experience.

Performance Support for Far-Transfer Tasks

In Chapter 3, I mentioned the use of electronic performance support systems (EPSS) designed to aid workers in procedural tasks. EPSS is designed to be used at the time of need in the workplace. Various forms of EPSS would also be useful to guide learners after far-transfer training. For example, for learning a sales engagement process, templates can serve as planning guides, sample proposals can illustrate best practices, and "quick tips" online reminders can refresh guidelines and their best applications. It's a good idea to consider the types of performance support that would best aid staff in the workplace, create them as part of the training design effort, and incorporate their use during the training to increase their likelihood of use after the training event

Evaluating Learning of Principle-Based Tasks

Evaluating the successful acquisition of principles is difficult because, as discussed above, there is typically no single set of "correct" answers. A range of responses that acceptably incorporate the guidelines must be identified. Test items should be similar to those provided during practice sessions. In general, they should include various job-realistic problems or scenarios that require the application of the target principles.

To help instructors score such tests consistently and accurately, use behavioral checklists. Figure 7.13 illustrates a partial checklist developed to train facilitators in effective techniques in virtual classroom settings. This checklist was used during both practice and testing sessions. Your instructors should be given training on the

Figure 7.13. Part of a Checklist for Evaluation of Virtual Classroom Interactivity

Copyright 2006, Clark Training & Consulting

 Virtual Classroom Program – Project 3 Interactions QA Form
Ruth Clark

1. The business objective and the overall presentation/ instructional goals for the VC program are:

Copy your objectives and goals from Project 1 Planning Form

To help presenters make most effective use of virtual classrooms in order to gain maximum cost-benefit from this delivery medium

To produce and deliver virtual classroom sessions that apply the DPEV model

2. Interactions occur frequently through the program:

o At least one or more per minute on average
o No presentations longer than two or three minutes without an interaction

 (YES)   NO

3. Type in slide title and underline the interaction type(s) and approaches used in each slide.

Slide Title	Interaction Type(s)	Interaction Approach(s)
1. Which features will your project use?	Demographic <u>Behavior</u> Attitude Knowledge Other None	Deductive Inductive <u>Lead In</u> Free Form Other
2. Which is better? Can graphics improve learning?	Demographic Behavior Attitude <u>Knowledge</u> Other None	Deductive Inductive <u>Lead In</u> Free Form Other

4. Treatments for Interactions—Quality Checklist
To be completed by instructor and reviewer partner

❑ Interactions are sufficiently frequent to sustain attention and promote program goals.

In some topics, more frequent interactions could be included. For example, all of the psychological functions were explained before any interactions were added.

❑ Interactions are interspersed throughout the program.

Yes, every topic has one or more interactions.

❑ Interactions are job relevant. They require participants to apply, not parrot back.

Yes, interactions require participants to apply guidelines and principles.

❑ Interactions use different types and approaches appropriate to the program goal and topic.

Many interactions are knowledge and lead in. More inductive types of questions could be used.

administration and scoring of far-transfer tests. During rater training, several instructors score sample performances of varying quality. Consistency of scoring can be calculated by inter-rater reliabilities, a mathematical procedure available in most test design handbooks. After each independent rating, the criteria checklist should be discussed until all instructors interpret the criteria in more or less the same way. Each successive rating session should produce higher reliability scores. Because of the greater judgment required to evaluate application of principles, passing scores should be more flexible than with near-transfer types of training.

COMING NEXT

Organizing Your Training

We have now looked at how to train all five types of content at the remember and use levels of performance and delivered either in classrooms or via e-learning. In Chapter 2, I provided some high-level tips for completing your job analysis and organizing your lessons. In the next chapter I will describe more detailed techniques for defining and organizing your technical content into courses and lessons. Then, in Chapter 9, I will look at some issues that uniquely apply to e-learning.

For More Information

Guided Discovery

Clark, R.C. (in press). *Building expertise*. San Francisco, CA: Pfeiffer. (See Chapters 1 and 11.)

Clark, R.C., & Mayer, R.E. (2007). *e-Learning and the science of instruction* (2nd ed.). San Francisco, CA: Pfeiffer.(See Chapter 14).

Worked Examples

Clark, R.C. & Mayer, R.E. (2007). *e-Learning and the science of instruction* (2nd ed.). San Francisco, CA: Pfeiffer. (See Chapter 10.)

Clark, R.C., Nguyen, F., & Sweller, J. (2006). *Efficiency in learning*. San Francisco, CA: Pfeiffer. (See Chapter 8.)

How to Organize Your Lessons and Exploit e-Learning Features

Chapter 8: Organizing Your Training Content

Provides more detail on how to define workforce learning content in a systematic way and how to organize that content into courses and lessons.

Chapter 9: e-Learning Design

Describes the unique features to be addressed in e-learning courses, including evidence-based guidelines for screen design, design of interactions and feedback, as well as use of audio and simulations.

Figure 8.1. Anatomy of a Lesson

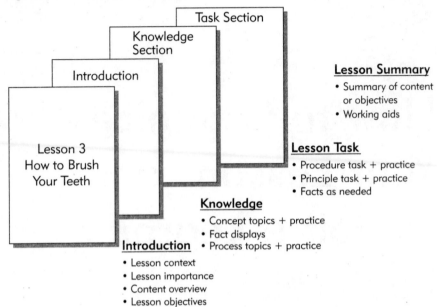

Task Section

Knowledge Section

Introduction

Lesson 3
How to Brush
Your Teeth

Lesson Summary
• Summary of content
  or objectives
• Working aids

Lesson Task
• Procedure task + practice
• Principle task + practice
• Facts as needed

Knowledge
• Concept topics + practice
• Fact displays
• Process topics + practice

Introduction
• Lesson context
• Lesson importance
• Content overview
• Lesson objectives
• Table of contents

8

Organizing Your Training Content

CHAPTER OVERVIEW

In the previous five chapters I have summarized the instructional methods that lead to learning of facts, concepts, processes, procedures, and principles at the remember and application levels. The introduction to structured lesson design in Chapter 2 suggested a generic model for lesson organization that includes four sections: a lesson introduction, a section on the knowledge associated with the task, the target lesson task, and a summary, as illustrated in Figure 8.1. In this chapter I will describe in greater detail how to define the content of your technical training program and, having defined it, how to organize it into lessons and a course.

Although I have saved the discussion of how to define and organize content for the end of this book, you would actually undertake this work during the early stages of the instructional development process, primarily during the task analysis phase. See Chapter 1 to review the instructional development process.

How to Define the Content of Your Training

Defining course and lesson sequences is one of the important early steps in packaging content into a form that allows it to be readily acquired by learners. There are five common pitfalls that I see associated with course and lesson planning. First, often content is organized around the structure of a product or the knowledge domain, rather than around the context of the job. For example, a course on how to use a new computer application will organize the lessons around application features such as menus and icons, rather than how workers will use the application on their jobs. Alternatively, a course for new supervisors will organize lessons around management theory and models, rather than supervisory tasks such as giving assignments and feedback. As a result, learning takes place but new knowledge is not organized in the brain based on how it will be used later on the job. A second common flaw is the inclusion of content that is nice to know merged with content that is essential to the job. The result is bloated training. Third, important supporting knowledge in the form of essential concepts or processes associated with the lesson procedural or strategic tasks is omitted. Often subject-matter experts are so familiar with the content that their unconscious competence leads to knowledge gaps in the training. Fourth, I often see courses where too much content—including essential content—is crammed into a single lesson. Learners get overloaded with too much content presented at once. The last common problem is a failure to include regular meaningful practice exercises. Often the learning objectives are too ambitious for the training time allotted. There is simply no time for practice. The focus shifts to content covered rather than skills learned. Unfortunately, "content covered" often fails to translate into skills learned.

Courses that display one or more of these flaws tend to be didactic and packed with technical content that is not always job-relevant. Learners can feel overwhelmed by a combination of missing information and too much information without practice. Many will assume the problems they experience reflect their own lack of competence and end up demoralized and unmotivated by the training experience. By applying the guidelines I summarize in this chapter, you can avoid these common traps. The first and foremost solution to many of these problems is to start your training effort with an organized and detailed job and task analysis.

Start with a Focus on the Job

The single most common and fatal problem I see in much course development today is lessons organized around theory or product features, rather than around job tasks. The result is a product-based or a knowledge-based sequence rather than a task-based sequence. For example, software training is typically organized around interface features such as menus and icons, rather than around how to apply software features to accomplish job tasks. Figure 8.1 compares a product-based approach to teaching Microsoft Word® to a job-based approach. Product or knowledge-based courses fail to transfer to the job. Since new knowledge and skills are not learned in the context of how they will be used on the job, learners have to struggle to translate what they learned in the class into something they can easily apply on the job.

The first rule to guide your training development efforts is: **Begin with the job.** Effective training for business and industry must improve work performance, and it does that best by teaching specific skills and knowledge needed to perform job-related tasks successfully. During your task analysis, you will identify the knowledge and skills used by best practitioners, as well as what the target population already knows. Prior knowledge, subtracted from the job-required knowledge, gives you the content of the training program.

Figure 8.1. Product-Based vs. Job-Based Software Training Outlines

Product-Based Outline		Job-Based Outline	
I	Pull Down Menu Bar Options: Overview	I	Functions of Microsoft Word: Overview
II	Use of Options Under File Pull Down	II	How to create, name, and save a short document
III	Use of Options Under Edit Pull Down	III	How to edit text: font, point size, style, positioning, spacing
IV	Use of Options Under View Pull Down	IV	How to do global search and replace
V	Use of Options Under Insert Pull Down	V	How to insert tables

Organize most of your lessons around job tasks (either procedures or principles). If you have many lessons in your program,
group them into larger segments—commonly called modules or
units. Define the anticipated outcomes of each lesson by writing
lesson objectives linked to the procedure or principle-based task that
is the focus of that lesson. Further define the content of each lesson
by outlining the introduction, knowledge, task, and review sections.
At the end of the process, you have a complete outline of each lesson in the course, as well as the lesson and unit learning objectives.

The job analysis technique presented here assumes you are
responsible for training an entire job. If you are only training a
subset of a job, your job analysis may focus on only one function
or a few tasks associated with a function. For example, if you were
designing customer service representative training for new hires,
you would be developing total job training. On the other hand,
if you were responsible for training employees on how to use the
automated interoffice mail and calendar system, you would be
developing training for several tasks such as scheduling meetings or
forwarding email that might be associated with a variety of job functions. In either case, you need to begin with the job and organize
your content around it.

This chapter has two major sections. The first describes a top-
down technique for doing a job analysis. The second illustrates several guidelines for transforming your job analysis data into course
and lesson outlines.

How to Do a Job Analysis

A top-down process for defining the knowledge and skills associated with job performance includes defining job functions, tasks,
guidelines or steps, and associated knowledge, as described in this
section. Although I am presenting this as a nice, neat top-down
analysis, I find that real life is much more chaotic. Often the
job analysis process may start from the top but tends to jump

around to various levels. Your goal at the end is to be able to fill in the major pieces described here.

1. Defining Job Functions

If you are training an entire job, begin by dividing the work into several non-overlapping functions. Job functions are similar to what your organization may call AORs (areas of responsibility), job duties, or KRAs (key results areas). In fact, you might be able to find a job analysis that was already done for the purpose of performance appraisal or selection testing. Check with your HR department to see whether a previous job analysis has been completed.

A job function defines a major responsibility resulting in a specific output that is relatively distinct from other outputs. For example, customer service representatives working for a major utility handle a variety of customer telephone calls. Since the entire universe of their job is handling telephone calls, an effective way to define the functions would be by types of calls. Observations of and interviews with customer service representatives and their supervisors reveal four major types of calls: (1) requests for service installations, (2) questions about bills, (3) requests for service repair, and (4) credit inquiries from customers who cannot pay their bills. Figure 8.2 summarizes the major functional areas for a supervisor. Note that each functional area leads to relatively independent job results.

Figure 8.2. Common Supervisory Job Functions

Selecting employees
Providing training
Making work assignments
Providing feedback on results
Aligning consequences with results
Managing employee problems

2. Defining Job Tasks Associated with Functions

Next look at each major job function (if you are training the entire job) and break it into several tasks. Defining job tasks is a somewhat subjective and usually iterative process. A job task is a defined set of specific steps or guidelines that result in a measurable outcome. When several tasks associated with a particular function are completed, that function is accomplished. Begin job-task definitions with a verb. There are a variety of ways to define the tasks associated with functions, none of them right or wrong. The major guideline is to manage the size of the tasks. Tasks that turn out to be very large result in lessons that are too lengthy. Tasks that are too small do not include sufficient content to make up a meaningful lesson. You will typically adjust and readjust your tasks as you find out more about them through the job analysis process. In general, shoot for tasks that include approximately five to fifteen steps or guidelines. I'll return to the size issue in the next section.

One of the functions of a supervisor is staff hiring. The five tasks associated with this function are documented on the Task Analysis Worksheet illustrated in Figure 8.3. Note that you need to decide whether the task is procedural or principle-based as you define each one. As with most supervisory tasks, the hiring tasks are all principle-based, meaning there will not be a single set of steps to follow. Instead, the hiring task is defined by guidelines that will have to be adapted by the supervisor to each situation.

3. Defining Steps or Guidelines Associated with Each Task

As you identify each of the major tasks associated with the functions, you will indicate whether it is near-transfer (based on a procedure) or far-transfer (based on principles or guidelines). If the task is primarily procedural, specify the steps that make up the procedure, as described in Chapter 3. If it is principle-based, define the guidelines, as described in Chapter 7. Remember that it's more challenging to derive guidelines than to specify steps. I summarized several approaches to job analysis of far-transfer tasks in Chapter 7.

The number of steps and guidelines will vary, but in general should not exceed a five-to-fifteen range in order to translate your tasks into brief lessons. These numbers are a general guideline; exceptions will depend on the experience of the target audience and

Figure 8.3. Tasks Associated with Hiring New Staff

TASK ANALYSIS WORKSHEET: TASKS OF FUNCTIONS

Training Program	Function	Page
Beginning Supervisors	Hiring New Staff	

Task#	Task	Description	Task Type (circle one)
1	Define staffing requirements	Based on projected projects and unit goals, define number and skills of new staff	Procedural / **(Principle-based)**
2	Prepare and submit job descriptions to HR	Use HR guidelines and templates to submit job descriptions	Procedural / **(Principle-based)**
3	Screen job applications	Identify those applicants that meet hiring specifications	Procedural / **(Principle-based)**
4	Write job interview guide	Prepare questions for job interview panel	**(Procedural)** / Principle-based
5	Conduct interviews and rate applicants	Conduct job interviews and rank applicants; submit to HR	Procedural / **(Principle-based)**
			Procedural / Principle-based
			Procedural / Principle-based

the need to keep task steps together for logical cohesiveness. But do work to limit the number of steps or guidelines per task so employees are not overwhelmed with new information. If your analysis results in thirty steps, the task is probably too large. Divide it into two tasks. This will keep your lessons short and manageable. Employees with more experience can generally manage longer lessons with more complex information, but short lessons are necessary for trainees who are new to the tasks being trained. Short lessons are especially important in e-learning to sustain attention.

In the supervisor job analysis shown in Figure 8.3, the fourth task is *Write job interview guide*. Some of the relevant guidelines based on HR documents and on observations of best practices among experienced supervisors are documented on the Task Analysis Worksheet illustrated in Figure 8.4. Note that each guideline begins with a verb and incorporates detailed job directions. If you write your steps or guidelines in a directive format during the job analysis, you can convert them quickly into tables or worksheets when you write your instructional materials.

With procedural tasks, write your steps in as specific a manner as possible. Thus, rather than writing for Step 1: "Determine whether the customer has prior service," a more specific action phrase is, Ask the customer, "Have you had service with Reliable Utilities prior to this time?" The rule of thumb is that trainees should be able to follow procedural steps without having to fill in many gaps. If you have omitted steps or the step size is too big for the target audience, they will get lost. Your gauging of the level of detail will depend in part on the background knowledge and experience of the audience.

As you can see on the bottom of Figure 8.4, it is at the task, level that you will define the major lesson objectives. The major lesson objective requires the learner to perform the task by following the steps or applying the guidelines presented in the lesson. You may also need to write a supporting objective if you have several important concepts linked to your guidelines.

4. Defining Associated Supporting Knowledge

Once you have broken tasks down into steps or guidelines, there is a final stage to complete the task analysis: identifying the knowledge associated with the steps or guidelines. The tasks and steps involve observable actions taken by employees; these are the how-to's of your training. However, associated with many steps or guidelines

Figure 8.4. Task Analysis Worksheet to Document Guidelines for Writing Job Interview Questions

TASK ANALYSIS WORKSHEET: PRINCIPLE–BASED TASKS

Training Program	Function	Task #	Audience	Page
Basic supervision	Hiring new staff	4	New supervisors	

Task Name: Write Job Interview Guide

Guidelines of the Task	Knowledge Needed	Content Type of Knowledge (circle one)	
Define required technical and performance skills needed for job.	What is a technical skill? What is a performance skill?	Fact	Concept
Write a question for each technical skill and for each performance skill.		F	C
Write questions that are job-relevant, open-ended, behavioral, and non-discriminatory.	What are job-relevant questions? What is an open ended question? What are behavioral questions? What are discriminatory questions?	F	C
Validate questions with HR	Hiring process at Acme House	F	C
		F	C
		F	C

Task/Lesson Objective (Terminal):

Given...	A staff hiring plan
the trainee will...	Write interview questions
at a standard of...	– match the required technical and performance skills – are job-relevant, open-ended, behavioral, and non-discriminatory

Knowledge Objective(s) (Enabling/Supporting):

Given...	Sample hiring questions
the trainee will...	Identify those that are job relevant, open-ended, behavioral, and non-discriminatory
at a standard of...	With no errors

are important facts or concepts that should be trained. For example, note that the fourth guideline in Figure 8.4 states: "Write questions that are job-relevant, open-ended, behavioral, and nondiscriminatory." To apply this guideline, the supervisor will need to know the concepts of job-relevant, open-ended, behavioral, and discriminatory questions. These items are documented in the right-hand column of the task worksheet.

Types of Supporting Knowledge

Most supporting knowledge involves concepts or facts. For example, ask yourself which of the following behaviors are needed:

- *Concept Discriminations.* Will they know it when they see it? Discriminations involve knowledge of concepts or facts. For example, for the guideline: *Write a nondiscriminatory question for the hiring interview,* the learner must be able to distinguish between questions that are and are not discriminatory. They would have to be able to apply the concept of discriminatory question.

- *Factual Information.* What specific, unique knowledge is associated with the steps or guidelines? Appropriate factual knowledge would be required to answer the following questions: Where is the equipment located? Where is Box 5 located on Form 3–19? What is my access code to log in?

Often job experts omit some of the important supporting knowledge associated with work tasks. To be sure you have defined all the key knowledge needed, look at each step or guideline in the task analysis and ask yourself what concepts or facts should be included. If there are several important knowledge items that are conceptual, you may write a supporting lesson objective that describes what learners can do to demonstrate they have acquired them. The supporting objective written at the bottom of the worksheet in Figure 8.4 links to the knowledge topics associated with the hiring interview task.

Defining Steps, Guidelines, and Knowledge: Knowing Your Audience

How small to make your step or guideline and what to include as knowledge depends both on the job to be trained and on the background knowledge of your potential audience. Prior knowledge and skills should be designated as prerequisites, and prerequisites are not taught. In contrast, knowledge topics are the underlying facts and concepts linked to the steps or guidelines that your trainees don't yet know. They have to be included in the training program.

To decide whether the job knowledge and skills are prerequisites or knowledge needed, you need to get a good idea of what your target audience already knows. You can interview sample target trainees or even test them to define their entry knowledge. If your trainee audience has varied background knowledge, you have three main alternatives:

1. *Design training for the lowest entry level of knowledge and skills.* If possible, provide introductory-level training separately for those trainees requiring it. Your course can be divided into introductory and advanced sections targeted toward the different entry-level skills of your trainees.

2. *Prepare pre-course work for entry-level trainees.* Pre-course work can sometimes be used to equalize background knowledge. For example, as part of a training program on calculation of customer bills for experienced customer service representatives, all representatives were asked to bring their calculators. Pilot course sessions revealed that many of the representatives could not effectively use the calculators. Therefore, an online pre-course assignment required each participant to complete and submit several mathematical problems using a calculator before attending class.

3. *Consider self-instructional training via e-learning.* Asynchronous e-learning is much easier to tailor to individual background differences than instructor-led training. Questions and tests can be used to assess the knowledge and skill

needs of different audience backgrounds, with dynamic matching of content to individual needs. Learners can complete a pretest at the start of the course and receive advice as to which lessons are most appropriate to them.

What About Processes?

As I discussed in Chapter 6, understanding and application of process knowledge is increasingly important for any kind of work that benefits from an understanding of systems. As you are identifying the tasks along with their steps or guidelines, probe for any related process information. Possibly the individual tasks are embedded in a larger business process. Or perhaps, if the tasks involve troubleshooting equipment, there are related equipment processes that would be helpful. Such process knowledge may be quite small and could be incorporated into the knowledge section of an individual lesson. Alternatively, it may be large and detailed and require a full lesson itself. Figure 8.5 shows a task analysis worksheet for documenting processes. The sample illustrates the hiring process—an important context-setter for new supervisors. Use the process worksheet to document relevant process stages when this knowledge will significantly impact the quality of task outcomes.

From Job Analysis to Course and Lesson Outlines

Once you have analyzed the job and target population and defined the functions, tasks, steps, guidelines or associated knowledge, including any related processes, you are ready to develop course and lesson outlines. Sequencing and combining the information defined in the job analysis into a course outline is a far-transfer task; there are no invariant steps. This section summarizes four principles to consider when developing your course outline: zoom, common-skills-first, spiral, and job-centered.

Figure 8.5. Task Analysis Worksheet Documenting Hiring Process

 TASK ANALYSIS WORKSHEET:
STAGES OF PROCESSES

Training Program	Function	Task #	Audience	Page
Basic supervision	Hiring new staff	4	New supervisors	

Process Name: The Acme Hiring Process

Stage	Description (What Happens)
1	Yearly unit goals and budgets are posted
2	Division labor requirements and hiring needs are defined
3	Division supervisors and HR develop job recruitment statements
4	HR recruits and screens applicants
5	Supervisors and HR develop and conduct hiring interviews
6	HR makes job offers

Knowledge Objective (Optional):

Given...	A complaint from a job applicant
the trainee will...	Define which stages may be the source of the complaint
at a standard of...	Stage activity and complaint content are congruent

The Zoom Principle

It is especially important for employees who are new to the information to get a broad overview of the content before getting the details. The novice needs to build a general mental framework on which to attach or "hang" the lesson details. Once you have presented the big picture followed by some relevant detail, return to the big picture periodically to orient the learner and emphasize interrelationships. Reigeluth (1983), uses a camera analogy to illustrate this principle. Begin the course with a wide-angle view capturing a broad picture of the content. Then zoom in on some detail. Before moving to a different detailed scene, bring the lens back out to the whole picture. This principle is especially important when your learners are new to the information and when your instruction will be delivered via computer.

Applying the Zoom Principle

You can implement the zoom principle by starting with a process lesson or with an overview of content. If you were teaching maintenance of the components of the erylitizer, you might begin with an overview of the glopple generation process, relating it to the internal equipment structures. If you were teaching the lesson on how to prepare a hiring interview, you might start with an overview of the hiring process to orient new supervisors.

A second way to implement the zoom principle is to begin the course with an overview of each major function or task to be trained. The new customer service representative should be presented with a general idea of the major types of calls he will be taking and how they relate to each other. Then, as the course progresses to detailed information on each call type, the big picture can be presented again periodically, giving even more detail on the relationships among call types. If you were teaching employees how to use a new automated system, a broad picture of the major software functions could serve as a recurring focal point, illustrating the relationships of specific screens to the entire system.

Some of the visual capabilities available in e-learning might lend themselves especially well to a zoom approach. For example in Figure 8.6 you can see a schematic diagram of how an AIDS virus invades a cell. Some areas in the cell schematic allow the learner to visually zoom in and get details on that stage of the process. See Chapter 6 for more details on this example.

The Common-Skills-First/Spiral Principles

These two principles complement each other. As you conduct your job analysis, you will find certain guidelines, procedures, concepts,

Figure 8.6. A High-Level View of the AIDS Process

Accessed from www.roche-hiv.com, january 4, 2007

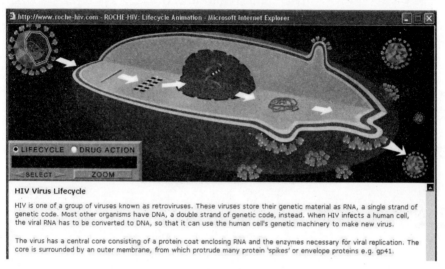

or facts associated with multiple functions or tasks. For example, the customer service representatives use the telephone and computer in connection with every job function. Identify such recurring "foundation" content and introduce it early in the course. For example, during the first week of call center customer service training, three basic skills are introduced: basic mechanics of the computer, mechanics of the telephone, and telephone communication skills. Later, each of these foundation skills is revisited at an increasing level of difficulty, as specific types of calls are trained. The iterative revisiting of introductory skills reflects the spiral principle.

The spiral principle, illustrated in Figure 8.7, suggests that courses in general should not follow a strictly linear sequence. Instead, train the foundation skills at an introductory level early in the course and build on those skills and in more advanced lessons. Telephone courtesy provides a good example. It would be a mistake to teach customer service during the first week of a six-week training program and never mention it again. Instead, introduce basic good customer service skills in the beginning. Then, as each type of customer interaction is taught, revisit how the service basics apply to

Figure 8.7. The Spiral Principle of Course Organization

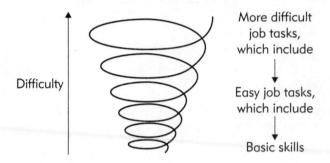

THE SPIRAL CURRICULUM

Difficulty

More difficult job tasks, which include

Easy job tasks, which include

Basic skills

that interaction. The spiral principle recommends that a critical skill be presented at a general level early and reinforced at greater levels of complexity throughout the course.

Three Organizational Frameworks

In addition to the principles described above, sequence and cluster your content by applying one of the following three organizational schemes (which are not mutually exclusive): logical prerequisite, job-centered, or problem-based. Let's look briefly at each.

Logical Prerequisite Sequences

Some instructional content has very strong logical relationships in the knowledge itself. Mathematics presents a good example. Suppose you are teaching ways to solve simple linear equations in algebra. Before solving equations, the students must master addition and multiplication of negative numbers. Content with these types of strong internal prerequisites may dictate some of your topic and lesson sequences. Skills A and B must be mastered before Skill C. Alternatively, you may have content for which, from a logical point of view, there are a number of sequencing alternatives that would work. In those cases, you can let one of the other frameworks guide your primarily organizational scheme.

Job-Centered Sequences vs. Knowledge-Centered

A knowledge-centered sequence organizes content around the structure of the content itself. From high school through university, most educational courses are organized this way. A typical history survey course will follow a chronological sequence starting with ancient and progressing to medieval and modern periods. A biology course will devote sections to zoology, botany, genetics, physiology, and other topics, based on the scientific domain. These knowledge taxonomies serve as the organizing force behind course structures. Often technical experts develop a complex and unique knowledge structure in their memories that may be taxonomic in nature. They will naturally tend to organize their courses following their own mental models. In software courses, a topic-based organizational scheme translates into feature-based training focusing on the menus and software features. While your educational background may suggest a knowledge-centered approach, my advice is: Don't do it! Use a job-centered approach instead.

A job-centered organizational scheme will group knowledge according to how it will be used on the job. The tasks that make up jobs serve as the basis for the sequence and clustering of lesson material. For example, if teaching Microsoft Word®, lessons would focus on job-related tasks such as creating and saving a short document, performing basic edits, cutting and pasting between documents, doing global search and replace, inserting tables and graphics, and so forth.

For most workforce learning, a knowledge-based structure is counter-productive! New knowledge and skills—even important knowledge and skills—when learned outside the context of their job application often will not transfer. Learners may get an "A" in the class. But later, back on the job, they are unable to apply what they learned. Instead, follow the task analysis process to define important job tasks and organize your content around those tasks.

Problem-Based Learning (PBL)

An adaptation of job-based learning uses problems or case studies drawn from the job as the organizing and driving framework for instruction. Each lesson uses a case study as the basis for learning the relevant knowledge and skills needed to solve the case. Unlike the prerequisite sequence described previously, the course organization is more holistic, with the case serving as the focal point for building the knowledge and skills needed. To design a PBL course, identify a sequence of varied cases so that, as learning progresses, all major knowledge and skill areas have been included. In addition, early cases should introduce more basic knowledge and skills, with later cases providing a context to review the earlier knowledge (spiral principle) and add new skills. For more information on problem-based learning, see the sources listed at the end of the chapter.

To sum up, as you group and sequence the skills and knowledge you identified in your job analysis, apply:

- The zoom principle to keep trainees oriented to the big picture

- The common-skills-first and spiral principles to move from easier foundation knowledge to more specialized and difficult content while revisiting the earlier-trained skills

- The job-based or problem-based principle to group most lesson content around work tasks related to improved organizational performance

A Sample Course Structure

Figure 8.8 illustrates a model course structure in which introductory training units that present the common skills are followed first by units to teach the easier job functions, then units to teach the more difficult ones. Units are made up of lessons that teach the important tasks linked to each function. The lesson is the fundamental organizational unit of all training programs. In the next section, I'll review some guidelines for organization of individual lessons.

Figure 8.8. Translating Functions and Tasks into Units and Lessons

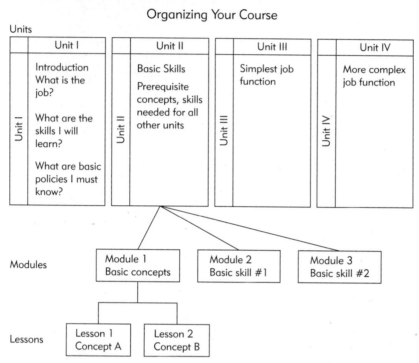

Organizing Your Course

Organizing Instructive Lessons

Instructive lessons use a step-by-step building-block approach. As a general rule, instructive lessons include an introduction, a section that includes major concepts or process topics identified as supporting knowledge. If you have several knowledge topics, end the knowledge section with a practice exercise. Follow the knowledge section with the lesson-task steps or guidelines, including a demonstration and practice. End the outline with a brief summary. In classroom courses, practice exercises are essential after the major task section. If the knowledge section includes several important topics, practice should be inserted there as well. Figure 8.9 illustrates an outline for a text lesson "How to Construct a Hiring Interview." The amount of detail to include in the outline will depend on who will read the

Figure 8.9. Sample Instructive Lesson Outline

Lesson Outline Template

Lesson Title: How to Write a Job Interview Guide

I. Lesson Introduction

Introduction Transition Statement	* Previous lesson on screening job applicants * This lesson focuses on preparing the interview guide
Importance (WIFM)	* Company cost to hire * Legal issues involved with interviews
Overview (Tell 'em what gonna tell 'em)	* Technical and performance skills * Questions that are job relevant, open-ended, behavioral, and non-discriminatory * Writing the interview guide
Objectives (Tell 'em what they will do)	* Identify technical and performance skills * Label effective and ineffective hiring questions * Write an interview guide
TOC	TBD

II. Knowledge Section

Topic 1	Review of hiring process
Topic 2	Identify technical skills and performance skills for a job
Topic 3	Job-relevant questions Open vs. close-ended questions Behavioral questions
Topic 4	Non-discriminatory questions
Knowledge Practice	Give sample questions for hiring cases and identify those that exemplify the topics above

III. Task Section: How to

List steps or guidelines or reference JTA worksheet	See guidelines on Task Analysis Worksheet

| Demonstration | Complete a whole-class participatory demonstration of developing an interview guide |
| Task Practice (Hands-on) | Assign teams to construct interview guide for case study jobs Project: Each new supervisor write a job description for a job on his or her team |

IV. Lesson Summary

Tell 'em what told 'em	A table summarizing the features of effective questions
Final Lesson Practice (optional)	NA
Job Aid Description	Legal guidelines Table of effective questions features
Quiz Description	Sample questions to match to case study job requirement Write five questions for a case-study job description

outline and your training development standards. In my example, I have taken a "lean" approach with sufficient information to indicate a sequence of topics that will need to be expanded in the student manual.

e-Lessons that follow a traditional "instructive" format should also include the general sections of introduction, knowledge needed, major tasks, and summary. But a need for very frequent interactions recommends that you place practice exercises after every major chunk of knowledge. How large a chunk you present between interactions depends on the experience of your audience. Novices need relatively smaller chunks with very frequent practice opportunities to give them a chance to build new knowledge gradually.

e-Lesson outlines may look similar to those for texts, although to accommodate greater flexibility and movement within the courseware, flowcharts may be used in addition to or instead of outlines. The expository screens present information using the content-specific display guidelines described in previous chapters. The practice screens include interactions that require trainees to process the information at the application level. In general, you should place interaction screens after every idea chunk presented in the information screens.

Outlines for Problem-Based Learning Lessons (PBL)

PBL lessons begin with and are driven by job case studies or problems. They typically allow much greater flexibility for the learner to access related information and to solve the case problem. Often, learning combines team sessions with asynchronous individual work. For example, a team may review a case, define a plan of attack, and then allow each individual to work one aspect of the case or to derive a solution to be compared with the solutions of other team members. Unlike the linear outlines illustrated in the previous section, PBL outlines begin with a definition of the major components of the case, as summarized in Figure 8.10, which shows the five major components for a PBL lesson on the hiring interview. All of the case components support the major task deliverable, which in this example is an effective interview guide. To enable learners to complete their case assignments, the lesson will typically include a trigger event, all of the case-relevant data the learner would need, such as job descriptions and data, feedback on case products, and instructional and reference resources learners can use to solve the case.

PBL lessons are most appropriate for more advanced learners who can handle greater flexibility and self-directed work to solve case problems. Also, PBL environments often benefit from collaboration—either synchronous or asynchronous. PBL lessons are better adapted to training of far-transfer tasks, for which there may be multiple different solutions. See resources at the end of this chapter to learn more about PBL learning environments.

Figure 8.10. A Template for a PBL Lesson Outline

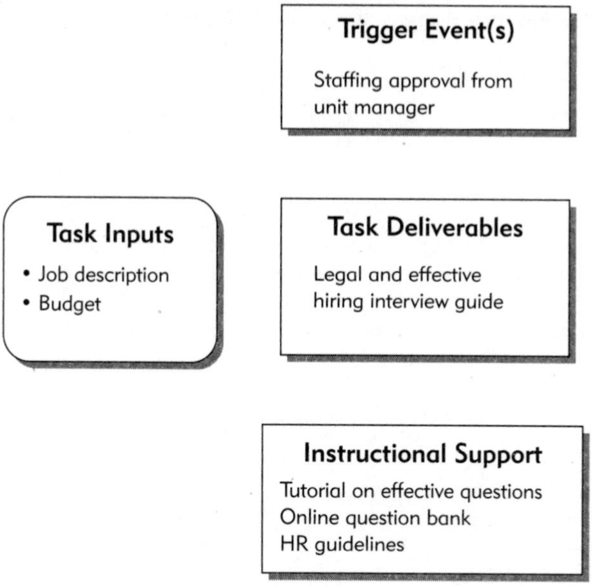

Trigger Event(s)

Staffing approval from
unit manager

Task Inputs

• Job description
• Budget

Task Deliverables

Legal and effective
hiring interview guide

Feedback

Peer and instructor
Review of questions

Instructional Support

Tutorial on effective questions
Online question bank
HR guidelines

Moving from Outline to Learner Materials

Once the outline has been completed and approved, the instructional team will use it as the basis for writing student manuals, creating exercises, drafting storyboards, creating job aids and tests, and all other components of the learning environment. The knowledge and task sections will be filled in with the information displays and practice exercises I described in the previous chapters. The lesson introduction will take the ideas listed in the outline and expand on them in the student manual and the instructor guide. In general, I recommend providing student notes that are lean but that also contain the core content of the lesson as well as directions to practice exercises. In the next sections, I summarize how to expand on the outline to write the student materials.

The Introduction

The lesson introduction is brief, but plays an important role in orienting and motivating the learner. In technical training, we often

Figure 8.11. Introduction Section from Hiring Interview Lesson

	How to Write a Job Interview Guide
Introduction	In previous lessons you have learned to define staffing needs and to screen job applicants. This lesson focuses on preparing for the job interview.
Importance	Your interview is your best opportunity to assess potential new hires. And preparation for the interview is essential for success. By writing your questions ahead, you avoid legal traps of discriminatory questions and craft questions that will elicit the best information for selection.
Overview	You will learn to develop questions to assess candidates' technical and performance skills. This lesson will describe the features of questions that are: o Job-relevant o Open-ended o Behavioral, and o Non-discriminatory
Objectives	You will be given sample staffing requirements and write an interview guide for the hiring interview. Your questions will be: o Job-relevant o Open-ended o Behavioral, and o Non-discriminatory

Table of Contents

Topics	Page
The Acme Hiring Process: Review	65
Technical vs. Performance Skills	66
Practice: Technical vs. Performance Skills	67
Job-Relevant Behavioral Questions	68
Practice: Job-Relevant Behavioral Questions	70
Non-Discriminatory Questions	71
Practice: Non-Discriminatory Questions	73
How to Write the Interview Guide	74
Practice: Write an Interview Guide	78

forget to sell the benefits of learning achievements. Use the lesson introduction as a marketing vehicle to promote the knowledge and skills to be gained. Figure 8.11 shows the classroom-lesson introductory page for the hiring interview guide lesson outlined in Figure 8.9. An effective lesson *introduction* orients learners by explaining why they are learning this particular information, in terms of what they have just learned and what they are going to learn. An *importance section* is designed to motivate by reinforcing the value of the lesson skills. The *lesson overview* gives a brief, high-level summary of the content, applying the zoom principle, while the *lesson objective* gives a clear statement of the actions the learner will take to demonstrate acquisition of the content. Although most lesson introductions are fairly brief—contained in a single page or two or three screens—they play a critical role by setting the stage for the lesson.

Knowledge Section

The next section of the lesson should teach all major concepts and processes identified in the task analysis. Depending on whether the key knowledge is a concept, process, or fact, follow the formats described in Chapters 4, 5, and 6 to display the required information and design practice. Research that compared teaching lesson concepts prior to the task with teaching them during the task found that teaching them prior to the task resulted in better learning. Teaching the major knowledge items ahead of the major lesson task will offload the mental work of trying to perform a new task and master all the associated information at the same time. For example, it will be easier for trainees to practice writing an interview guide once they already know the types of questions that are most appropriate.

Lesson Task Section

The third section teaches the major tasks to be learned. Follow the guidelines in Chapters 3 through 7 for teaching procedures or principles with relevant practice exercises.

Lesson Summary

Last, a summary should briefly encapsulate the major points covered and perhaps preview the material in the lesson to come.

Check Your Understanding

To practice writing a lesson outline, try the short exercise in the Appendix for Chapter 8.

Design Documents

If your training program is of sufficient size, criticality, or cost, it is a good idea to put together a course blueprint, commonly called a design document, after you have completed the task analysis and lesson outlines. The design document summarizes the decisions you have made on the four ingredients of instruction—content, objectives, methods, and media. It also documents relevant project-management information. The purpose of the design document is to get all stakeholders to agree on what is to be included, what outcomes can be expected, and all project details, *before* starting the development phase of your effort. Typical design documents include the course outlines, major instructional objectives, delivery plans, timelines, and required resources, and instructional treatment sections. A sample lesson may also be included. e-Learning documentation may also include storyboards for one lesson, flowcharts, or a prototype functional lesson.

COMING NEXT

e-Learning Design

Up to this point, I have emphasized the similarities between development of e-learning and classroom learning environments. However, there are some major differences to consider when designing e-learning. Chapter 9 will look at issues to consider, including:

- Synchronous versus asynchronous forms of e-learning
- Design of screens versus pages
- Use of audio and video media
- Design of online interactions and feedback
- Use of simulations and games

For More Information

Clark, R.C., & Kwinn, A. (2007). *The new virtual classroom.* San Francisco, CA: Pfeiffer. (See Chapter 11 on problem-based synchronous e-learning.)

Clark, R.C., & Mayer, R.E. (2007). *e-Learning and the science of instruction* (2nd ed.). San Francisco, CA: Pfeiffer. (See Chapter 14 on problem-based e-learning.)

Jonassen, D.H. (2004). *Learning to solve problems: An instructional design guide.* San Francisco, CA: Jossey-Bass.

Mayer, R.E. (Ed.). (2005). *The Cambridge handbook of multimedia learning.* New York: Cambridge University Press.

Figure 9.1. e-Learning and Classroom Training Converge

Based on data from Sugrue and Rivera, 2005

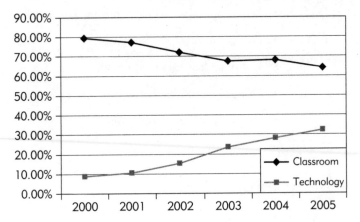

9

e-Learning Design

CHAPTER OVERVIEW

In Chapter 1 I described delivery media as one of the four fundamental ingredients of all training programs. As I mentioned in Chapter 1, it is not the delivery medium that causes learning but rather the instructional methods I've reviewed in Chapters 3 through 7. However, not all media are equivalent! In this chapter I will introduce some key factors to consider in design of e-learning environments that are different from classroom instruction. Of course, entire books have been written on e-learning—including my two books, *e-Learning and the Science of Instruction* with Dr. Richard Mayer and *The New Virtual Classroom* with Dr. Ann Kwinn. I include a few recent resources at the end of this chapter. My goal in this chapter is to introduce some of the major issues that are unique to either synchronous or asynchronous digital learning environments, including:

- Screens versus pages

- Audio versus text

- Interactions and feedback

- Digital simulations and games

Digital Versus Classroom Learning

As you can see in Figure 9.1, the proportion of face-to-face classroom and various forms of electronic learning delivery environments is approximating a 50-50 split. Recent data on training media usage summarized in Figure 9.2 show that synchronous (virtual classroom) and asynchronous (self-study) forms of e-learning are just about even, at around 15 percent each.

While savings in travel costs and training time are the major drivers of the growing e-learning market share, there are some instructional drivers as well. Although learning from either a classroom or an e-learning lesson that are designed using the same instructional methods is more or less equivalent, not all media can deliver the same instructional methods. For example, books are limited to text and a few graphics, compared to computers, which can provide both static and dynamic visuals, audio, interactions with feedback, and simulations. The question is not which delivery medium is better. The better question is: Which mix of media is best to achieve your learning goals? As you plan a learning *process* rather than a learning event, incorporate delivery media that can

Figure 9.1. e-Learning and Classroom Training Converge
Based on data from Sugrue and Rivera, 2005

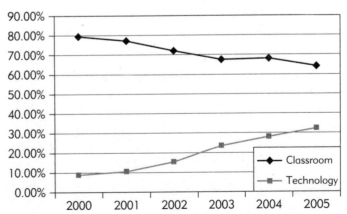

Figure 9.2. Percentages of Delivery Media Used in Organizational Training

Based on data from *Training*'s 2006 Industry Report

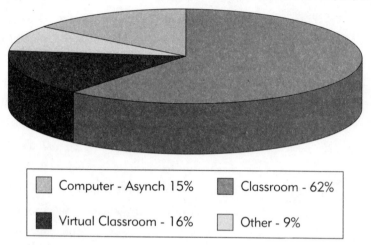

| Computer - Asynch 15% | Classroom - 62% |
| Virtual Classroom - 16% | Other - 9% |

be accessed at diverse points in the learning continuum—some at a time and place convenient to the learner, some in a training setting, some accessible from the work setting, and some that allow individual reflection as well as others that support collaborative work.

Whether you are planning for synchronous, asynchronous, or both types of e-learning, your challenge is to harness the features of each delivery medium to maximize learning. When considering e-learning, you need to address the following unique features:

- Design of screens rather than pages
- Use of modalities of visuals, audio, and text
- Timing and features of interactions with feedback
- Potential for simulations and games

I will provide a few quick tips in this chapter, and I encourage you to find out more in the resources provided at the end of the chapter.

Engaging Learners Through Screens

In any form of e-learning, your main interface with the learner is the screen rather than the classroom instructor. One of the most important elements to consider as you move to e-learning is content visualization. Because many of us have worked in word-dominant media such as workbooks, texts, and instructor guides for so long, transitioning to a more visual medium can be a challenge. When designing e-learning, we need to think more like video producers than textbook producers. Compare Figures 9.3 and 9.4 that show before and after student projects in my virtual classroom design course. Note that the first version screen (Figure 9.3) relies too heavily on text and small visuals. In the improved version (Figure 9.4), most of the text is converted to instructor audio and the visuals have been enlarged. Adding relevant visuals to explanatory text has consistently shown learning improvements averaging around 90 percent. But what exactly constitutes an effective visual?

Figure 9.3. A Virtual Classroom Screen with Too Much Tiny Text and Small Visuals

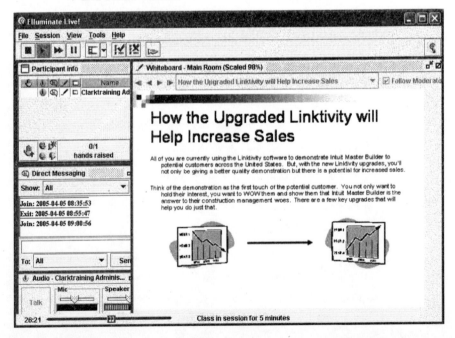

Figure 9.4. An Improved Screen with Text Converted to Audio and Visuals Enlarged

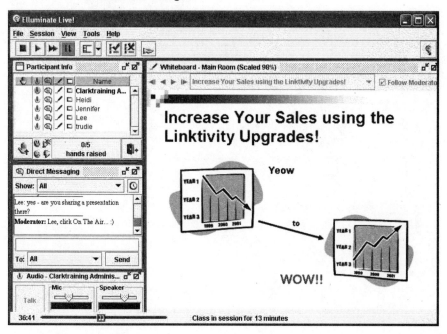

Visuals for Learning

Evaluations of school textbook graphics reported that over 80 percent of the visuals served no useful instructional purpose. This dismal finding reflects an over-reliance on decorative visuals instead of more meaningful visuals that illustrate relationships among content elements. An over-reliance on clip art can lead to visuals that are pretty but do not promote learning. In fact, an excessive use of decorative visuals has been shown to depress learning. For example, the space thematic visual treatment shown in Figure 9.5 is an attempt to make a computer application course more engaging. In fact, it runs the risk of impeding learning. Instead, consider ways to use explanatory visuals that show qualitative or quantitative relationships, illustrate changes in time or space, or make intangible content concrete. For example, Figure 9.6 shows a transformational visual that illustrates a step-by-step demonstration of a computerized telephone

Figure 9.5. A Fantasy Thematic Visual Does Not Promote Learning.

From *Graphics for Learning*, by Clark and Lyons, 2004

management procedure in an asynchronous e-learning lesson. For more information on visualization of content, see my book, *Graphics for Learning* co-authored with Chopeta Lyons.

Apply the Four Cs

As you start to plan your screen layouts, attend to the four Cs: Clean, Consistent, Clear, and Controlled by the learner. You can see the four Cs in Figure 9.6. This asynchronous e-learning screen is not overly cluttered with information, and white space plays a predominant role in the layout. The placement of key elements such as the computer interface, the call recipient, and the callers is consistent throughout the lessons. Clarity of your visual message is supported

Figure 9.6. A Transformational Visual Illustrates Procedural Steps
From e-Learning and the Science of Instruction, Clark and Mayer, 2007

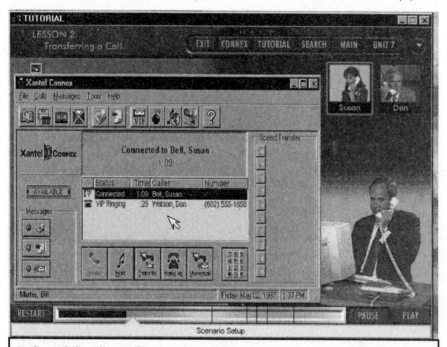

Audio: While Bill is talking to Don, Julie calls with a question. Bill knows that Julie needs to talk to Sally in the Art Department and decides to transfer her while he is talking to Don.

by placement of your visuals. More important visual information should be placed in the more dominant areas of the screen. Note in Figure 9.6 that the computer interface is allocated the predominant center screen real estate, while the telephone respondent and callers are located in less predominant areas of the screen. The bar along the bottom of the screen indicates audio progress. The navigation options are clearly located at the top right-hand portion of the screen. Always provide consistent and easily accessible access to exit, menu, forward, and backward buttons. In the virtual classroom, similar guidelines apply except that screens typically progress under the control of the instructor rather than the learner.

Audio and Text

In both synchronous and asynchronous e-learning, you can explain relevant visuals with either audio narration or with on-screen text. When should you use audio to present words? and When is text better?

Recent research has shown that a complex visual such as the one in Figure 9.6 is better explained by audio narration than by on-screen text. The reason is that working memory has two processing centers: one for visual information and one for phonetic or auditory information. When you explain a complex visual with text, you overload the visual center with two visual inputs: a graphic and the text. Mental overload leads to depressed learning. Static visuals with many elements or almost any dynamic visual are examples of complex visuals. Because animations compress so much visual information into a short time frame, they are almost always classified as complex.

However, reliance on audio alone can pose a problem for learners who may lack sound cards or have hearing disabilities. Therefore, in asynchronous e-learning, offer an "audio-off" option that will replace audio with on-screen text. In synchronous forms of e-learning, many tools offer a closed captioning facility that can be accessed by those needing it. Closed captioning is also a good option when the audio narration is in a different language from the learner's native language.

It's a common myth that learning is better when a visual is explained by BOTH audio and on-screen text. The truth, however, is that presenting the same words in audio and text overloads working memory and reduces learning. Present words with audio or with text, but not both.

While research evidence recommends audio to explain complex visuals, sometimes on-screen text is a better option. Audio is transient. By that I mean that, once it's played, it's gone. That's why you should always provide a replay button. However, for any situation in which learners will need to refer back to words or have access to words over a period of time, they should be placed on the screen in text. A common example is directions to a practice exercise. Text directions should explain what the learner is supposed to do, including specifically how they should respond. Some examples include: "Review the sample web page screens and place a star on those that meet corporate standards" (virtual classroom), or "Drag each sample web page that 'meets' or 'fails' corporate standards to the correct bin. You can click on each sample web page to see an enlargement" (asynchronous).

Digital Interactions

In Chapters 3 through 7, I summarized the main type of practice exercises you can use to promote learning of each of the five content types. Those fundamental methods apply to classroom as well as to digital learning environments. However, there are some guidelines you should consider that are unique to e-learning.

1. Build in Frequent Interactions

In either synchronous or asynchronous e-learning environments, learner attention spans are a major challenge. In e-learning, usually taken at the learner's desktop, it's easy for participants to minimize the screen and attend to the myriad of other potential distracters in the workplace. The main tool you have to sustain attention and learning is frequent meaningful practice interactions throughout your lessons. Whereas in a typical classroom session you may not schedule a practice episode for as long as twenty or thirty minutes, in e-learning your interactions should be scheduled much more frequently—as often as every minute! Naturally, including so many more interactions will require more instructional time, meaning that you may not be able to "cover" as much content as in a typical face-to-face classroom. Just remember that materials "covered" is not the same as content learned, regardless of the delivery medium.

2. Vary Response Options

Most e-learning tools offer a variety of response formats. For example, in the virtual classroom you can ask learners to type into chat, mark the whiteboard, select a survey option, or speak through the microphone. In asynchronous e-learning, learners can click on-screen objects, type short text statements on the screen, select multiple-choice options, or engage in simulated learning environments. Although we tend to pose question options in text, consider also visual options. For example, in Figure 9.7, learners are asked to use polling to vote for the visual they felt was most effective.

3. Provide Explanatory Feedback

Research on feedback shows that it's important to let learners know when they are correct or incorrect as well as to give an explanation. In asynchronous forms of e-learning, feedback to multiple-option questions appears on the screen as the learner selects an option.

Figure 9.7. Visual Response Options in a Virtual Classroom Session on Graphics

From *The New Virtual Classroom*, Clark and Kwinn, 2007

Which Is Better...
This?   Or This??

A   **B**

To give feedback to more open-ended responses such as typing in a short phrase, you can post some model answers for the learners to compare with their responses. In the virtual classroom, the instructor typically provides feedback with a combination of verbal explanations as well as visual illustrations that reveal the correct answer.

In addition to traditional instructional feedback, you can also consider using what I call "intrinsic" or naturalistic feedback. This type of feedback is most often included in asynchronous simulations. In intrinsic feedback, the learner responds in a simulation and then experiences the results of his or her choices. For example, in Figure 9.8 I show a screen shot from a course on customer service. The learner selects one of three dialog responses to a customer comment. Each response results in a different reaction from the customer—exhibited in the customer's body language and in her verbal comments. In this example, the customer's reactions are an example of intrinsic feedback. In addition, instructional feedback is

---

Figure 9.8. A Customer Service Simulation Offers Intrinsic and Instructional Feedback

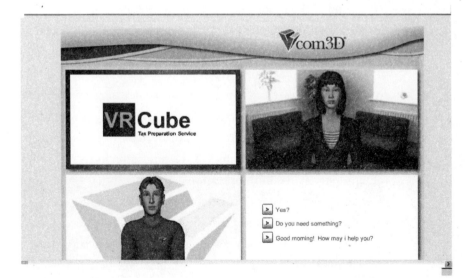

provided by the learning agent located in the lower-left-hand side of the screen.

Games offer many examples of naturalistic feedback. In an arcade game, a poorly steered vehicle crashes into environmental obstacles. In a learning simulation, actions taken lead to real-world types of consequences. For example, if the learner selects a suboptimal approach, a sale is lost, a customer is unhappy, or project costs go up.

Simulations and Games in e-Learning

The popularity of online games makes the use of simulations and games in e-learning an intriguing possibility. As with many topics in this chapter, entire books are written on electronic simulations and games. Here I offer a brief introduction and invite those interested to read more in the references provided at the end of the chapter.

A simulation is a model of a real-world environment that is rule-based. That is, when the user takes an action, the electronic environment responds in a real-world manner. There are two main types of instructional simulations: operational and conceptual. An operational simulation is designed primarily to teach procedures. For example, the lesson shown in Figure 9.6 includes opportunities for learners to practice using the computerized phone system by clicking on the telephone control options on the simulated interface screen. Research has shown that operational simulations do effectively teach skills that can be emulated in an e-learning environment. Operational simulations are especially effective ways to practice skills that may be dangerous, infrequent, or difficult to practice in the workplace. For example, Figure 9.8 is a simulation focusing on customer service skills. Both operational and conceptual simulations offer an opportunity for learners to practice over and over again until they can perform automatically.

Conceptual simulations have been used in the educational arena to teach science concepts. For example, in a biology lesson learners can manipulate genes on chromosomes and see changes in the features of an imaginary creature. In a physics lesson, learners can raise the temperature or pressure and see changes in the volume of a gas. In the commercial sector, simulations are used as learning environments to teach principle-based tasks such as loan analysis, sales skills, patient diagnosis and treatment, equipment troubleshooting, computer network analysis, workplace safety, and customer service skills, to name a few. Figure 9.9 shows a screen from a simulation in which the learner can access various office tools to gain information about a bank loan applicant. For example, a credit report can be requested via fax, books provide industry background information, and the learner can even "visit and interview" the applicant. After accessing relevant information, the learner makes a loan decision by clicking an "approve" or "reject" button and typing in a justification for his or her decision.

Figure 9.9. A Bank Loan Simulation

With Permission from Moody's Financial Services

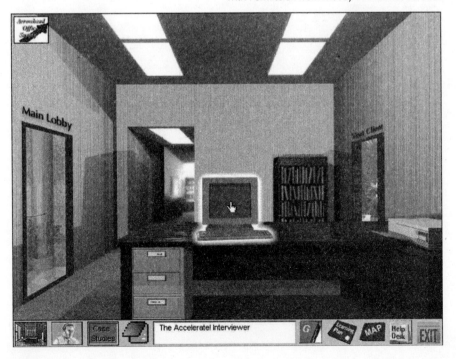

Conceptual simulations are more challenging to develop than operational simulations. To avoid the inefficiencies of random exploration of a conceptual learning environment, embed learner support such as an on-screen agent, a repository of worked-out examples, templates with guidelines, or even traditional tutorial lessons. In addition, a series of case problems should start with simpler situations and progress to more complex scenarios. Research on troubleshooting simulations showed an acceleration of expertise in which, after twenty-five hours on the simulation, two-year technicians demonstrated the same competencies as ten-year veterans. The acceleration of expertise comes from an opportunity to gain experience from simulated situations that in the real world would consume a great deal of elapsed time. For example, in real-world troubleshooting, equipment failures may be rare and may not reflect the full range of potential problems. A simulation offers an opportunity to experience an "ideal" sequence of equipment failures in a compressed time period.

Online Games and Learning

Because the term "games" spans such a diverse spectrum, it's difficult to make any generalizations about the learning value of games. Learning will depend on how effectively a game is aligned to the learning goals. All too often, training courses over-use game models such as Jeopardy that only promote memorization of facts or concepts.

To qualify as a game, an online event must offer the learner a challenge to achieve a goal, respond on the basis of game rules, offer an interactive high-engagement experience, and incorporate consequences. Factors reported to add appeal to games include a fantasy element in the interface, an optimal level of challenge (not too easy or too difficult), and control over the progress or results of the game. Some of the more common categories of games include arcade games (also called twitch games) that reward speed and accuracy of responses, game-show environments that usually reward accuracy of recall or recognition of facts or concepts, and strategy or adventure games that may involve a story line and reward decisions that reflect the use of guidelines that are the focus of learning.

Research on games is just beginning. However, even at these early stages we know that the learning effectiveness of a game will depend on how well game features align to the instructional goal. The Oregon Trail® is a game that failed instructionally—at least in its initial implementation—due to a mismatch of game features to learning goals. The objective of the game was to help fifth through seventh graders learn to solve problems faced by pioneers in wagon trains. Two aspects of the wagon-train game proved counter-productive to learning. Players were allowed to shoot animals as a food source and also to move the wagon trains over diverse terrain. However, players co-opted these features by turning the simulation into an arcade "shoot 'em up" game as well as a racing game. If your goal is one that requires higher levels of thinking or problem solving, avoid "twitch" elements that will defeat the reflection needed for learning.

Another important feature of games is instructive feedback to guide learning. For example, a botany game in which learners constructed plants best suited to survive in imaginary planetary environments yielded best learning when an on-screen learning agent commented on players' choices (Moreno, 2004). For example, when an incorrect choice was made, the agent would respond: "Hmmmm. Your deep roots will not help your plant collect the scarce rain that is on the surface of the soil. Here is a better choice. . . ."

Finally, manage the complexity of your game or simulation. Comparisons of simulation interfaces with high- and low-complex elements have found better learning with less complexity. For more information on games and simulations, see Chapter 15 in *e-Learning and the Science of Instruction* (2nd ed.).

Designing for e-Learning

When comparing learning from various types of digital environments with that from face-to-face classrooms, we consistently find that what causes learning is not the delivery medium per se but how effectively the features of any delivery medium are used to promote learning. We have all been to classroom sessions that were ineffective, and poor e-learning environments are easy to find. The instructional methods presented throughout this book apply to classroom and electronic learning environments. The main differences relate to visualization of content and interactivity. Digital learning requires you to engage learners to sustain attention and promote learning through relevant visuals and frequent interactions. In addition, exploit the features included in e-learning tools and interfaces such as audio, video, simulations, and games in ways that lead to learning.

With each edition of this book, I see great evolutions in e-learning tools and opportunities. For example, in the previous edition, synchronous e-learning was in its infancy. Likewise, in the past asynchronous e-learning has required specialized programming skills. New tools and techniques are making the mechanics of e-learning development easy. Therefore, it is more important than ever to exploit the unique features of e-learning in ways that are compatible with human learning processes. As electronic learning resources continue to evolve, you will find new ways to implement the basic instructional methods summarized in this book.

For More Information

Clark, R.C., & Kwinn, A. (2007). *The new virtual classroom.* San Francisco, CA: Pfeiffer.

Clark, R.C., & Lyons, C. (2004). *Graphics for learning.* San Francisco, CA: Pfeiffer.

Clark, R.C., & Mayer, R.E. (2007). *e-Learning and the science of instruction* (3rd ed.). San Francisco, CA: Pfeiffer.

Clark, R.C., Nguyen, F., & Sweller, J. (2006). *Efficiency in learning.* San Francisco, CA: Pfeiffer.

Horton, W. (2006). *e-Learning by design.* San Francisco, CA: Pfeiffer.

Mayer, R.E. (Ed.). (2005). *The Cambridge handbook of multimedia learning.* New York: Cambridge University Press.

Moreno, R. (2004). Decreasing cognitive load for novice students: Effects of explanatory versus corrective feedback in discovery-based multimedia. *Instructional Science, 32,* 99–113.

Exercise for Chapter 1
The Four Ingredients of Training

Read the brief instructional development scenario below and list the information or content, the performance outcome, the instructional methods, and instructional media for this program.

Diane plans to teach a refresher class to experienced sales representatives at Optical Scientific Products on how to best close a sale. She spends a month observing and interviewing effective and less effective account representatives, noting similarities and differences in how they handle their sales calls. She summarizes five guidelines used by the best representatives that are not used by the less effective performers.

Diane summarizes the five guidelines in a sales support website to be used during and after the training for reference. Then she videotapes several examples of the guidelines being applied in various customer situations to be used in an asynchronous e-learning program to illustrate different approaches. Diane writes out some role-play scenarios for classroom practice to be administered and critiqued by the instructors and by some consultant clients.

Exercise for Chapter 3
Distinguish Between Linear and Decision Procedures

For each procedure listed below, indicate whether it is a linear or a decision procedure:

1. How to repair the malfunctioning copy machine

2. How to calculate the standard deviation

3. How to approve or disapprove a customer application

4. How to determine which tax form to use for different clients

5. How to compute a customer's car insurance rate

Exercise for Chapter 4
Identifying Concepts

For each step or guideline listed below, underline the concepts included:

1. Enter all normal and overtime worked into the B31 column.

2. Divide the numerator by the denominator.

3. Count the number of defects and defectives.

4. Write the gross adjusted income into column B4.

5. Avoid discriminatory questions in your behavioral interview.

Exercise for Chapter 5
Distinguishing Between Facts and Concepts

The steps and guidelines below contain both facts and concepts. Indicate whether the italicized content should be treated as a fact or a concept.

1. Use the *F4 key* to bring up the customer calculation formula.

2. Multiply the *radius* by Π.

3. Disengage the *alpha demodulator* first.

4. Ask the customers whether they have had *prior service* with our company.

5. Enter all normal time worked in section 4 of the *company timesheet.*

6. Enter *your password* in the *sign-on field.*

Exercise for Chapter 6
Identifying Processes

For each example below, indicate whether it reflects a procedural task or a process. If it's a process, does it refer to a technical or a business process?

1. How the university evaluates and admits new students

2. How to complete your university application form

3. How to troubleshoot the XP3400 printer

4. How the XP3400 printer works

5. How to enter a query into the search software

6. How search software uses relevance feedback to rank documents

Exercise for Chapter 7
Distinguishing Between Procedural (Near-Transfer) and Principle-Based (Far-Transfer) Tasks

For each of the following tasks, indicate whether the task is more procedural or more principle-based.

1. How to develop an effective hiring interview

2. How to complete the company timesheet

3. How to calculate standard deviation

4. How to design a web page

5. How to construct and refine an effective search query

Exercise for Chapter 8

Outlining a Lesson

Use the task analysis data below and create a lesson outline.

Task: How to develop an Excel spreadsheet for expense tracking

Step	Action	Knowledge
1.	Select File, New	Where are these options?
2.	Plan and label column headings	Columns vs. rows
3.	Set up page layout for summary sheet	What are page layout options? What is a summary sheet?
4.	Set up cell formats and enter formulas	What is a cell? What is a formula? What are formatting options? What are formula operators?
5.	Link cells that require updating	What is a link?

Exercise Solutions

Chapter 1

The Four Ingredients of Training

Information or Content: The five guidelines for closing a sale

Performance Outcome: Given a series of simulated account-representative customer interactions and the website reference, the learners will apply the guidelines for closing a sale so that their body language and words are congruent with the behavioral checklist.

Instructional Methods: The list of five guidelines; examples of applying the guidelines; role-play—apply the guidelines in customer scenarios

Instructional Media: Instructor, video, website, computer

Comments: The *media* deliver the instructional methods. In this situation the computer and video will deliver the examples, the website the list of five steps, and the instructor the practice and feedback. The *instructional methods* are techniques used to cause learning. In this situation the five guidelines are presented on the website supplemented by examples that illustrate their application. The practice methods are role play with feedback. The *content, or information,* of the program is the five guidelines needed to close the sale, derived through observation of expert performance.

The *performance outcome* states what the learners will do at the end of the class to demonstrate they have acquired the content. In this case, they will need to apply the five guidelines to the close of a sale.

Chapter 3

Distinguish Between Linear and Decision Procedures

1. How to repair the malfunctioning copy machine
   Decision (assumes if. . . . Then logic)

2. How to calculate the standard deviation
   Linear

3. How to approve or disapprove a customer application
   Decision (assumes if. . . . Then logic)

4. How to determine which tax form to use for different clients
   Decision (assumes if. . . . Then logic)

5. How to compute the customer's car insurance rate
   Both—will likely embed decision steps (assumes different rates depending on customer criteria) into an overall linear procedure

Chapter 4

Identifying Concepts

1. Enter all <u>normal</u> and <u>overtime</u> worked onto the B31 column.

2. Divide the <u>numerator</u> by the <u>denominator</u>.

3. Count the number of <u>defects</u> and <u>defectives</u>.

4. Write the <u>gross adjusted income</u> into <u>column</u> B4.

5. Avoid <u>discriminatory questions</u> in your <u>behavioral interview</u>.

Comments: Note that concepts are classes or categories for which you could provide multiple examples. Note that while column is a concept, the B31 column is very specific information and will be considered a fact—not a concept—as will be discussed in the next chapter.

Chapter 5

Distinguishing Between Facts and Concepts

1. Use the *F4 key* to bring up the customer calculation formula.
   Fact (a unique specific key)

2. Multiply the *radius* by Π.
   (radius = **Concept**; Π = **Fact**—a specific unique number)

3. Disengage the *alpha demodulator* first
   Fact (a unique specific part of equipment)

4. Ask the customers if they have had *prior service* with our company
   Concept (many examples exist)

5. Enter all normal time worked in section 4 of the *company timesheet*
   Fact (unique specific form)

6. Enter *your password* in the *sign-on field* Both are
   Facts, since they represent unique and specific information.

Comment: Ask yourself, can I give a definition and are there multiple instances of this item? If yes, you are likely dealing with a concept such as radius or prior service. On the other hand, if the item is unique, specific, or every instance is just like every other instance, the example is a fact such as Π, which is a unique specific number or a specific part of a form or equipment.

Chapter 6

Identifying Processes

1. How the university evaluates and admits new students
   Process; business

2. How to complete your university application form
   Procedure

3. How to troubleshoot the XP3400 printer
   Procedure

4. How the XP3400 printer works
   Process; technical

5. How to enter a query into the search software
   Procedure

6. How search software uses relevance feedback to rank documents
   Process; technical

Comments: Note that procedures are directive and are written in active voice, while processes are descriptive and written in passive voice.

Chapter 7

Distinguishing Between Procedural (Near-Transfer) and Principle-Based (Far-Transfer) Tasks

1. How to develop an effective hiring interview
   Principle-based (guidelines would lead to multiple correct solutions)

2. How to complete the company timesheet
   Procedure (step-by-step task—no judgment involved)

3. How to calculate standard deviation
   Procedure (step-by-step task)

4. How to design a web page
   Principle-based (involves application of guidelines; no single correct solution; requires judgment)

5. How to construct and refine an effective search query
   Principle-based (guidelines could be applied in different ways for a range of possible solutions)

Comments: Many tasks can be approached as a procedure or a principle. When not sure, ask yourself whether:

1. The task would be completed more or less the same way each time it is done.

2. Completing the task involves a minimum of judgment.

3. Consistency in task completion is important.

These are all indicators of procedures. In contrast, a principle-based task is indicated when:

1. The task has no one set of steps to follow each time.

2. The outcomes of the task rely on circumstances and judgment.

3. The outcomes are not necessarily consistent; multiple solutions exist.

Chapter 8

Outline a Lesson

I. Introductory Page(s)

Introduction. Make transition statement from previous lesson(s).

Importance. Give statistics on budget overages in last year; relate to need to carefully track expenditures and time savings using Excel.

Overview. List major lesson topics:

Objective(s): You will develop an Excel spreadsheet to track expenditures ensuring company expenses do not exceed the budget amount. You will use the XYZ authorized budget list and the PTN computer system.

You will distinguish between a summary sheet and a committed sheet and a formula and a link.

You will identify cells and formula formatting options.

Order of Topics:

Introduction
Summary Sheets vs. Committed Sheets
Rows vs. Columns and Associated Formats
Cells, Formulas, and Associated Formats
What Are Links
How to Develop the Spreadsheet

II. Lesson Body

A. Knowledge Topics   This lesson will include the following knowledge topics:

- Summary Sheets vs. Committed Sheets
- Rows vs. Columns and Associated Formats

- Cells, Formulas, and Associated Formats
- Links

B. Knowledge Practice   Learner will be given several completed spreadsheets and identify specific sections to include: type of sheet, rows, columns, cells, formulas, and links.

Learners will be given calculation goals and sample formulas and asked to select the appropriate formula to accomplish the goal.

C. Lesson Task   Learners will follow along on a step-action table as they view a demonstration of how to accomplish an Excel calculation.

D. Task Practice   Learners will be assigned several budget goals to complete by setting up the appropriate spreadsheet in Excel.

E. Lesson Review   A working aid will summarize the major formatting rules and steps to set up the spreadsheet.

GLOSSARY

Abstract Concepts

Content that involves a category that is intangible and cannot usually be represented with a diagram or picture. Some examples include integrity and empathy.

Action Table

A two- or three-column table commonly used to display procedures in manuals. Typical column heads are "Step" and "Action."

Analogy

An explanation used to help learners understand a concept, process, or procedure. Analogies typically are drawn from a different domain than the lesson content but share some common features with the target content. For example, a slice of pie is used as an analogy for the concept of a fraction.

Analysis

A part of the instructional systems design process during which the performance factors, including training, needed to reach operational objectives are defined. See also performance analysis and training needs analysis.

Animation

A series of still graphics that are projected rapidly to illustrate motion. Movie cartoons are a familiar form of animation.

Application Level

One of two levels of psychological processes that can be applied to content. At the application level, learners are able to use the content in a manner similar to the way it will be used on the job. Contrast with *remember level*.

Application Sharing   A virtual classroom facility that allows the instructor to show all participants a software application running on his or her computer. Often used to demonstrate computer applications in software training.

Architecture   A course or lesson design plan that specifies the organization and placement of course content, the amount and placement of learner interactions, and the degree of learner control over sequencing. Two examples include directive (also called instructive) and guided discovery.

Assessment   A measure of learning. Typically takes the form of a test.

Association Statements   A type of factual information in which multiple concepts are related to each other in a sentence. For example, "There are seven men in this room."

Asynchronous e-Learning   Instructional programs delivered on a computer that are designed primarily for self-study. These programs can be taken at any time by any one at his or her own pace. Common examples include web-based training courses.

Automaticity   A skill that is encoded into long-term memory and can be deployed without using working memory resources. Automated skills can be completed quickly and error-free without conscious processing. Automaticity builds only after hundreds of practice trials.

Blended Learning   The use of multiple instructional delivery media that may allow for a combination of instructional methods, for example, in self-paced self-study modes and in group collaborative modes.

Breakout Room   A virtual classroom facility that allows small groups of participants to work together independently with access to all of the facilities available in the main room.

Business Process   A flow of events or activities that involve multiple employees, departments, or organizational units. Some examples include a hiring process and performance appraisal process. Contrast with *technical and scientific process*.

Chat   A virtual classroom facility that allows all participants to type comments into the interface. Depending on the options selected, messages can be seen by all or can be sent to selected individuals.

Cognitive Load   The amount of mental work imposed on working memory. Cognitive load can come from the difficulty of the instructional content, the manner in which instructional materials are organized and displayed, and the mental processes required to achieve the learning goal.

Coherence Principle   Learning is better when extraneous visuals, text, or audio are omitted from instruction rather than included. See also *Las Vegas instruction*.

Collaborative Learning   A structured instructional interaction among two or more learners to achieve a learning goal or complete an assignment.

Common Skills First   An organizational principle recommending that knowledge and skills needed to complete many job tasks are sequenced at the start of a training event and then revisited as they are applied specifically to different tasks in more advanced lessons. See also *spiral principle*.

Communication Function   The purpose of a graphic to help learners build the correct mental model. For example, a procedure is conveyed by a transformational graphic that shows a series of steps either with arrows and text in a line drawing or with an animated demonstration.

Communication Modes   See *modes*.

Completion Examples   Worked examples in which some steps are demonstrated by the instruction and the remainder are filled in by the learner.

Concepts   Lesson content that involves a group of objects, events, or symbols called by the same name. Some examples are chair, spreadsheet, and virtual classroom. See also *concrete concept* and *abstract concept*.

Conceptual Simulation   A simulation in which the interface responds based on cause-and-effect rules. Typical examples include a genetics simulation or a simulation of ideal gas laws. Often used to teach scientific principles.

Concrete Concepts   Content that involves a group of objects that have parts and boundaries and can be represented with a diagram or picture.

Concrete Facts   Content that refers to a unique and specific object with parts and boundaries. A specific computer screen interface is a common example.

Content-Performance Matrix   A two-dimensional educational taxonomy that defines five types of content that can be processed at two levels.

Critical Incident   A story that illustrates how an important challenge or task was accomplished by an experienced performer. Used as a technique to define guidelines for far transfer or principle-based tasks.

Data   A common type of factual information that
                                 includes specific and discrete quantitative
                                 or qualitative information about objects.
                                 New product specifications are one common
                                 example.

Decision Procedure   A routine task that requires a different sequence
                                 of steps based on decision points embedded in
                                 the procedure. For example, if A, then do Steps 1
                                 through 3; if B, then do steps 4 through 6.

Decision Table   A two- or three-column table commonly used to
                                 display decision procedures in manuals. Typical
                                 column heads are "If" and Then."

Decorative Visual   A graphic used to add aesthetic appeal or humor.

Definition   A succinct statement of the critical features of a
                                 concept class.

Delivery Media   See *media*.

Demonstration   A step-by-step illustration of how to perform
                                 a task. May be provided by an instructor, on a
                                 video, or an animated e-learning sequence.

Design   A part of the instructional systems design process
                                 during which data from a job analysis are used
                                 to define course objectives and outline learning
                                 content.

Design Document   A report that summarizes the main outputs
                                 of the design phase of the project. Typically
                                 includes results of the job analysis, course
                                 and lesson outlines, learning objectives, and a
                                 summary of the instructional methods planned.

Development   A part of the instructional systems design process
                                 during which the learning materials, including
                                 student manuals, e-learning screens, practice
                                 exercises, video examples, etc., are created.

Diagram   An iconic representation of content.

Directive Architectures   Lessons or courses that are highly structured and
                                typically incorporate short learning episodes with
                                brief explanations, examples, and practice with
                                feedback. See also *instructive events*.

Discrimination   A psychological activity in which one item is
                                distinguished from other items. For example,
                                defective parts are discriminated from parts that
                                meet specifications.

Drill-and-Practice   A type of interaction that involves repetitive
                                responses to questions with the goal of fast and
                                accurate response. In earlier times, flash cards
                                were used for drill and practice.

e-Learning   Learning environments mediated by a computer.
                                These include (1) asynchronous self-paced
                                courses (formally called computer-based
                                training), (2) synchronous real-time instructor-
                                led courses (also called virtual classrooms), as
                                well as (3) blends such as a recorded synchronous
                                course reviewed asynchronously.

Electronic Performance   A digitally available guide intended to help
   Support   learners perform job tasks on the job.

Evaluation   A part of the instructional systems design process
                                during which the effectiveness of the training
                                is measured. Commonly defined in four levels:
                                learner reaction, learning, job transfer, and
                                bottom-line impact.

Evidence-Based Practice   The incorporation of valid research evidence
                                into decisions about selection or design of
                                instructional programs. The research is usually
                                based on controlled group experiments.

Example   An instance of a concept class. Examples are
                                an essential instructional method for teaching
                                concepts. Contrast with *non-example*.

Explanatory Feedback   Information given to learners in response to an action taken during the training telling them whether their actions were correct or incorrect and giving the reasons for the correct or incorrect assessment. For example: "Sorry, that is incorrect. You should have divided the numerator by the denominator."

Explanatory Visual   A graphic that illustrates relationships among content and helps learners build deeper understanding. Some examples include graphic expressions of quantitative data such as bar charts, relationships among qualitative data such as concept maps, and interpretive visuals such as schematic diagrams.

Exploratory Architecture   A learning environment in which learners are allowed high degrees of learner control, such as selecting which lessons to review, which exercises to complete, or which instructional methods to access. The Internet is an example of a large exploratory environment.

Facts   Lesson content that refers to unique, specific one-of-a-kind objects, events, or symbols. Some examples include specific application screens, codes, and forms.

Faded Worked Examples   A demonstration in which some steps are worked by the instruction and the remaining steps are worked by the learner.

Far Transfer   Knowledge and skills that must be applied in diverse contexts. Some examples include making a sales presentation or designing training. To perform far-transfer tasks, the worker must use judgment to adapt guidelines to diverse situations. See also *principle-based tasks*.

Flow Diagram	A chart typically including visuals, boxes, and arrows that depicts process stages.
Functions	The highest level of job responsibility defined in a job analysis. Also known as duties or key result areas. Functions are often listed on a performance appraisal.
Game	An online event that offers the learner a challenge to achieve a goal, responds on the basis of game rules, offers an interactive high engagement experience, and incorporates consequences.
Guided Discovery Architectures	Course or lesson design plans in which participants derive knowledge and skills from study of multiple examples, problem or task assignments, or other experiential activities. These courses rely on an inductive model of learning. See also *problem-based learning*.
Guideline	The smallest action statement associated with a principle-based or far transfer task.
Inductive Learning	Opportunities for participants to derive guidelines, features, or rules from viewing examples or engaging in experiences, for example, asking participants to list the features of a good website after viewing samples of well-designed and poorly designed websites.
Instructional Content	The information that makes up the knowledge and skills workers need for job success. Includes facts, concepts, processes, procedures, and principles.
Instructional Methods	The psychologically active ingredients of any learning environment that determine the learning effectiveness of that environment.

	Some examples include examples, practice assignments, and feedback.
Instructional Mode	See *mode*.
Instructional Systems Design	A systematic process for the planning, creation, and evaluation of learning environments that includes stages associated with design, development, and evaluation of training.
Instructional Technology	A field that focuses on the design and development of training programs that meet organizational objectives.
Instructive Events	Opportunities for participants to learn through highly guided instructional environments that typically provide explanations, examples, practice, and feedback. See *directive architecture*.
Instructive Learning	A training design in which key content is explained and illustrated followed by practice exercises. Most traditional training applies an instructive approach.
Integrated Content	Organization of lessons in which supporting concepts are presented at the same time as lesson tasks or process stages. Contrast with *segmented content*.
Interface	The design of an application screen that offers end users various functionalities such as buttons, icons, and windows. Effective interfaces are intuitive, simple, and user-friendly.
Interpretive Visual	A graphic used to illustrate an invisible theory, principle, or cause-and-effect relationship. An equipment schematic and a molecular model are two examples. Also called interpretive graphic.
Intrinsic Feedback	Feedback that comes from real-life consequences of a performer's response. For example, when

hitting a tennis ball, the sound and movement of the ball provide intrinsic feedback. Contrast to *extrinsic feedback.*

Job-Centered Sequences   An organizational framework in which lessons are organized around job tasks rather than theory, product features, or domain structures. These sequences are recommended for most workforce training. Contrast with knowledge-centered sequences.

Knowledge-Centered Sequences   An organizational framework in which lessons are organized around knowledge domains rather than job tasks. Knowledge-centered structures are common in educational settings in which the main topics are based on the content domain.

Learning Agent   An on-screen character used often in asynchronous e-learning courses that offers various forms of explanations, advice, or feedback to support the learning process.

Learning Objectives   A specific statement of what learners will do to demonstrate they have achieved the goals of a lesson or course. Usually includes an action statement, a condition statement, and a criterion statement. Also called learning outcomes. See also *terminal and supporting learning objectives.*

Learning Outcomes   A specific statement of what learners will do to demonstrate they have achieved the goals of a learning experience. Also called *learning objectives.*

Linear Procedure   A routine task that is completed from Step 1 to Step x in the same sequence each time.

Logical Prerequisite Sequence   An organizational framework in which skills are sequenced so that basic skills needed to accomplish learning goals are taught first.

Long-Term Memory	A relatively permanent mental repository of knowledge and skills in the form of mental models that are the basis for expertise. The mental models in long-term memory interact directly with working memory to influence the virtual capacity of working memory.
Media	The delivery technology used to provide instructional materials to learners. Includes items such as books, video, and computers. Also called delivery media.
Mental Model	A knowledge structure stored in long-term memory that is the basis of expertise.
Methods	A psychologically active ingredient of a lesson or course that promotes learning. Some examples include examples, practice exercises, analogies, and feedback. Also called *instructional methods*.
Mnemonics	A memory aid in the form of a visual or auditory cue.
Modality Principle	A proven instructional principle stating that complex visuals are understood more efficiently when explanatory words are presented in an audio modality rather than when presented in a written modality. Because working memory includes separate processing areas for visual and auditory information, using the auditory mode along with the visual makes most efficient use of limited working memory resources.
Modes	Three fundamental elements for communicating new content and instructional methods. Includes text, graphics, and audio.
Naturalistic Feedback	Response to learner interactions that reflect real-world consequences. For example, if an incorrect

customer response is made, the sale is lost. See also *intrinsic feedback*.

Near Transfer   Knowledge and skills that are applied in more or less the same way each time they are used. Refers to routine tasks such as logging into email or starting an automobile. Also called procedural tasks.

Non-Example   An instance that is similar to but not a member of a concept class. Often non-examples are used to help learners avoid discrimination errors in concept learning. Also called counter-example or invalid example.

Operational Simulation   A simulation in which the learner performs procedural steps and the simulated system responds in ways similar to the actual system. Flight simulators and software simulations are typical examples. Contrast with conceptual simulation.

Organizational Visual   A graphic that shows qualitative relationships among content. A tree diagram is a common example. Also called organizational graphic.

Pacing   The rate at which information is delivered and the source of control of information delivery rate. Instructor-paced is typical of instructor-led events, whereas learner-paced is typical of asynchronous media such as books or asynchronous e-learning.

Performance Analysis   A part of the instructional systems design process during which factors that will close gaps between desired organizational outcomes and actual organizational outcomes are defined. Typical performance factors include knowledge and skills, job standards, aligned incentives, and better business processes.

Performance Support   Guidelines provided to workers on the job to help them accomplish tasks. Also called job aids. See also *electronic performance support*.

Performance Test   A structured evaluation experience during which the learner performs a hands-on task and is evaluated for proficiency based on a specified standard of performance.

Principle-Based Tasks   Tasks performed by adapting guidelines to varying contexts of the work environment. Some examples include making a sales presentation or designing training. These tasks require workers to use judgment to apply guidelines to new situations. Also called far-transfer tasks or strategic tasks.

**Problem-Based
Learning (PBL)**   A form of guided discovery in which a course or lesson begins with a work-relevant problem or task and learning is fostered as participants solve the problem or complete the assignment. PBL uses an inductive approach to learning.

Procedural Tasks   A task made up of steps that are performed more or less the same way each time. Procedures are also called near-transfer tasks. Contrast with *principle-based tasks*.

Process   A flow of activities that involves multiple entities, individuals, or organizations. A description of how things work. Some examples include how brakes work, the hiring process, and how blood circulates in the body. See business, technical, and scientific processes.

Receptive Architectures Learning environments characterized by dissemination of information with limited or no opportunities for overt learner participation. A typical lecture and a documentary video are two examples of receptive designs.

Redundancy Principle   Learning is more inefficient when sources of information are duplicates and hence risk overloading working memory. A common example is narration of on-screen text used to explain a complex visual. The audio and visual expression of words is redundant and is proven to depress learning.

Reference-Based Training   An instructional design in which the facts and procedural steps are documented in a reference guide that is then used as part of the training design. Typically, the training manual will refer learners to the reference guide to complete exercises.

Rehearsal   The mental processing of new information in working memory that results in the integration of new information with existing activated knowledge to form new mental models stored in long-term memory. Effective practice exercises lead to productive rehearsal.

Relational Visual   A graphic intended to summarize quantitative data, such as a bar chart or a pie chart.

Remember Level   One of two levels of psychological processes that can be applied to content. At the remember level, learners can recall or recognize the basic content elements. For training purposes, in most cases, the application level is more appropriate than the remember level. Contrast with *application level*.

Representational Visual   A graphic that is intended to depict an object. A screen capture or an equipment photograph are two examples.

Scientific Process   A flow of natural events such as how blood is circulated through the heart or the weather cycle.

Segmented Content	Organization of lessons in which supporting concepts are presented prior to lesson tasks or process stages. Contrast with *integrated content.*
Self-Explanation Questions	Questions inserted into examples that require learners to explain the example or elements in the example. Often used to promote learning from examples.
Simulation	An artificial system that responds like a real-world system. Digital simulations are commonly used to illustrate processes and principle-driven relationships.
Social Presence	The degree to which various media features mediate interpersonal communication and thus allow participants to feel connected to others. Face-to-face learning environments are high in potential for social presence compared to print media, which are relatively low in potential social presence.
Spiral Principle	An organizational guideline that recommends that technical training content be revisited throughout a course or lesson at increasingly more complex levels.
Split Attention	The dividing of limited mental resources among related content items or facilities that result in distraction or extra effort invested to integrate the materials.
Stage	The smallest element associated with a process. One element in a series that constitutes what happens.
Steps	The smallest action statement associated with a procedure.

Strategic Tasks   Work accomplishments that rely on guidelines
                                  workers must apply using their judgment
                                  regarding the specific situation. Sales is a
                                  common example of a strategic task. See also
                                  principle-based and far-transfer tasks.

Structured Notes   Handouts for learners that use structured writing
                                  to display training content.

Structured Observations   A job analysis technique in which workers are
                                  observed and their actions (and sometimes
                                  thoughts) are recorded in a systematic way.

Structured Writing   A method for the organization and display of
                                  words and visuals that leads to most effective
                                  access and retrieval of information.

Supporting Knowledge   Information usually in the form of concepts
                                  and facts learners must know to apply steps or
                                  guidelines in order to complete tasks.

**Supporting Learning
Objective**   A specific statement of what learners will do to
                                  demonstrate they have acquired the supporting
                                  knowledge elements of a lesson. Contrast with
                                  terminal learning objective.

Surface Features   The outward appearance or features of graphics,
                                  such as whether they are an animation or line
                                  drawing.

Synchronous e-Learning   Computer-mediated learning environments
                                  in which multiple geographically dispersed
                                  participants are engaged at the same time in an
                                  instructional event. Also called virtual classroom
                                  or remote live training.

Task   A job accomplishment that results in a specific
                                  measurable outcome. Tasks may be based
                                  on procedural steps or on principle-based
                                  guidelines. Tasks are often the basis for lessons in
                                  technical training programs.

Task Analysis	A process whereby a job is defined to identify the key knowledge and skills that must be included in a training program. Also called job analysis.
Task Deliverable	The desired end product and process of a problem-based learning lesson. Should mirror best-practice products and processes from the work environment.
Task Input	Data learners will need relevant to their problem assignment in problem-based learning to enable them to create the task deliverable. Also called case-supporting data. Should mirror the types of case data that would be accessible in the work environment.
Taxonomy	A classification system. This book is based on a taxonomy called the content-performance matrix.
Technical Process	A flow of events that involves physical objects such as equipment. Some examples include how a hydraulic system works and the electronic control process for equipment X.
Technical Training	Learning environments delivered in face-to-face classrooms or via computer designed to build job-relevant knowledge and skills that improve bottom-line organizational performance.
Terminal Learning Objective	A specific statement of what learners will do to demonstrate end-of-lesson or end-of-course goals. Contrast with *supporting learning objective*.
Transfer of Learning	The retrieval and application of new knowledge that has been stored in long-term memory during training later when needed in the workplace.

Training Needs Assessment	A part of the instructional systems design process during which the major knowledge and skills as well as learning objectives of a training program are defined. Typically occurs after a performance assessment defines a gap in knowledge and skills.
Transformational Visual	A graphic used to show changes or movement in objects over time or space. A demonstration of how to interact with computer software is a common example.
Trigger Event	An initiating assignment or event that introduces learners to a project or case at the start of a problem-based learning lesson.
Varied Context Worked Examples	Demonstrations that incorporate diverse job-related scenarios. For example, in sales examples would illustrate different customers and different products.
Virtual Classroom	Synchronous computer-mediated learning environments with facilities for visualization, instructor and participant audio, and participant responses via polling and chat, among others. Some commonly used virtual classroom tools include Breeze, Elluminate, Centra, Live Meetings, and WebEx. See also *synchronous e-learning*.
Whiteboard	A facility in the virtual classroom interface that allows projection of slides as well as drawing or typing by the instructor and participants.
WIFM	What's in it for me? A statement or illustration of the importance or relevance of a lesson typically included in the lesson introduction.
Worked Examples	A step-by-step demonstration of how to solve a problem or complete a task. See also *varied context worked examples*.

Worked Problem   A task analysis technique in which a job expert talks aloud as he or he solves a job problem related to competencies to be built in the training. The goal is to define the mental processes experts use during task execution.

Working Memory   A central element of human cognition responsible for active processing of data during thinking, problem solving, and learning. Working memory has a limited capacity and storage duration for information.

Zoom Principle   A training organizational guideline that recommends starting with a high-level overview, moving in to teach details, and ending with the high-level overview.

REFERENCES

Anderson, L.W., & Krathwohl (Eds.). (2001). *A taxonomy for learning, teaching, and assessing: A revision of Bloom's taxonomy of educational objectives.* New York: Longman.

Bloom, B.S. (1956). *Taxonomy of educational objectives, handbook I: The cognitive domain.* New York: David McKay.

Clark, R.E. (1994). Media will never influence learning. *Educational Technology Research and Development, 42* (2), 21–30.

Clark, R.C. (in press). *Building expertise* (3rd ed.). San Francisco, CA: Pfeiffer.

Clark, R.C., & Kwinn, A. (2007). *The new virtual classroom.* San Francisco, CA: Pfeiffer.

Clark, R.C., & Lyons, C. (2004). *Graphics for learning.* San Francisco, CA: Pfeiffer.

Clark, R.C., & Mayer, R.E. (2007). *e-Learning and the science of instruction* (2nd ed.). San Francisco, CA: Pfeiffer.

Clark, R.C., & Nguyen, R (in press). Behavioral, cognitive, and technological approaches to performance improvement. In J.M. Spector, M.D. Merrill, J.J.G. van Merriënboer, & M.P. Driscoll (Eds.), *Handbook of research on educational communications and technology* (3rd ed.) Mahwah, NJ: Lawrence Erlbaum Associates.

Clark, R.C., Nguyen, F. & Sweller, J, (2006). *Efficiency in learning.* San Francisco, CA: Pfeiffer.

Clark, R.E. (1994). Media will never influence learning. *Educational Technology Research and Development, 42,* 21–30.

Drucker, P.F. (1998). The coming of the new organization. *Harvard Business Review on Knowledge Management.* Boston, MA: Harvard Business School Publishing.

Gagne, R.M. (1985). *The conditions of learning and theory of instruction.* Austin, TX: Holt, Rinehart, and Winston.

Gagne, R.M., & Medsker, K.L. (1995). *The conditions of learning, training applications.* Belmont, CA: Wadsworth.

Gentner, D., Loewenstein, J., & Thompson, L. (2003). Learning and transfer: A general role for analogical encoding. *Journal of Educational Psychology, 95*(2), 393–408.

Gordon, J., & Zemke, R. (2000). The attack on ISD: Have we got instructional design all wrong? *Training, 37,* 43–53.

Gott, S.P., & Lesgold, A.M. (2000). Competence in the workplace: How cognitive performance models and situated instruction can accelerate skill acquisition. In R. Glaser (Ed.), *Advances in instructional psychology: Educational design and cognitive science.* Mahwah, NJ: Lawrence Erlbaum Associates.

Horton, W. (2006). *e-Learning by design.* San Francisco, CA: Pfeiffer.

Industry Report. (2006). *Training, 38*(12), 20–32. Accessed December 8, 2006, from www.Trainingmag.com.

Jonassen, D.H. (2004). *Learning to solve problems: An instructional design guide.* San Francisco, CA: Jossey-Bass.

Kirkpatrick, D. (1994). *Evaluating training programs: The four levels.* San Francisco, CA: Berrett-Koehler.

Krathwohl, D.R., Bloom, B.S., & Bertram, B.M. (1973). *Taxonomy of educational objectives, the classification of educational goals. Handbook II: Affective domain.* New York: David McKay.

Lajoie, S.P., Azevedo, R., & Fleiszer, D.M. (1998). Cognitive tools for assessment and learning in a high information flow environment. *Journal of Educational Computing Research, 18,* 205–235.

Mager, R.F. (1997). *Measuring instructional results* (3rd ed.). Atlanta, GA: Center for Effective Performance.

Mager, R.F. (1997). *Preparing instructional objectives* (3rd ed.). Atlanta, GA: Center for Effective Performance.

Mager, R.F., & Pipe, P. (1997). *Analyzing performance problems* (3rd ed.). Atlanta, GA: Center for Effective Performance.

Mayer, R.E. (Ed.). (2005). *The Cambridge handbook of multimedia learning.* New York: Cambridge University Press.

Mayer, R.E., Hegary, M., Mayer, S., and Campbell, J. (2005). When static media promote active learning: Annotated illustrations versus narrated animations in multimedia instruction. *Journal of Experimental Psychology: Applied, 11*(4), 256–265.

Mayer, R.E., Mathias, A., & Wetzell, K. (2002). Fostering understanding of multimedia messages through pretraining: Evidence for a two-stage theory of mental model construction. *Journal of Experimental Psychology: Applied, 8,* 147–154.

Merrill, D.M. (1983). Component display theory. In C. M. Reigeluth (Ed.), *Instructional design theories and models: An overview of their current status.* Mahwah, NJ: Lawrence Erlbaum Associates.

Moreno, R. (2004). Decreasing cognitive load for novice students: Effects of explanatory versus corrective feedback in discovery-based multimedia. *Instructional Science, 32,* 99–113.

Robinson, D.G., & Robinson, J.C. (1995). *Performance consulting.* San Francisco, CA: Berrett-Koehler.

Rothwell, W.J. (2004). *Mastering the instructional design process.* San Francisco, CA: Pfeiffer.

Rothwell, W.J. (2005). *Beyond training and development* (2nd ed.). New York: American Management Association.

Rossett, A. (1999). *First things fast: A handbook for performance analysis.* San Francisco: CA: Pfeiffer.

Rossett, A., & Schafer, L. (2007). *Job aids and performance support.* San Francisco, CA: Pfeiffer.

Shrock, S., & Coscarelli, W. (2000). *Criterion-referenced test development* (2nd ed.). Silver Spring, MD: International Society for Performance Improvement.

Skinner, B.F. (1961). Teaching machines. *Scientific American, 205*(11), 90–102.

Sugrue, B., & Rivera, R.J. (2005). *State of the industry.* Alexandria, VA: American Society for Training and Development.

Wolff, E.N. (2005). The growth of information workers in the U.S. economy. *Communications of the ACM, 48*(10), 37–42.

Zemke, R., & Kramlinger, T. (1982). *Figuring things out: A trainer's guide to needs and task analysis.* Reading, MA: Addison-Wesley.

Zemke, R., & Rossett, A. (2002). A hard look at ISD. *Training, 39*(2), 26–35.

INDEX

RUTH COLVIN CLARK, Ed.D., has focused her professional efforts on bridging the gap between academic research on instructional methods and application of that research by training and performance support professionals in corporate and government organizations. Dr. Clark has developed a number of seminars and has written six books, including *e-Learning and the Science of Instruction, Building Expertise,* and *Efficiency in Learning,* that translate important research programs into practitioner guidelines.

A science undergraduate, she completed her doctorate in instructional psychology/educational technology in 1988 at the University of Southern California. Dr. Clark is a past president of the International Society of Performance Improvement and a member of the American Educational Research Association. She was honored with the 2006 Thomas F. Gilbert Distinguished Professional Achievement Award by the International Society for Performance Improvement and is an invited *Training Legend* Speaker at the ASTD 2006 International Conference. Dr. Clark is currently a dual resident of Southwest Colorado and Phoenix, Arizona, and divides her professional time among speaking, teaching, and writing. For more information, consult her website at www.clarktraining.com.

Pfeiffer Publications Guide

This guide is designed to familiarize you with the various types of Pfeiffer publications. The formats section describes the various types of products that we publish; the methodologies section describes the many different ways that content might be provided within a product. We also provide a list of the topic areas in which we publish.

FORMATS

In addition to its extensive book-publishing program, Pfeiffer offers content in an array of formats, from fieldbooks for the practitioner to complete, ready-to-use training packages that support group learning.

FIELDBOOK Designed to provide information and guidance to practitioners in the midst of action. Most fieldbooks are companions to another, sometimes earlier, work, from which its ideas are derived; the fieldbook makes practical what was theoretical in the original text. Fieldbooks can certainly be read from cover to cover. More likely, though, you'll find yourself bouncing around following a particular theme, or dipping in as the mood, and the situation, dictate.

HANDBOOK A contributed volume of work on a single topic, comprising an eclectic mix of ideas, case studies, and best practices sourced by practitioners and experts in the field.

An editor or team of editors usually is appointed to seek out contributors and to evaluate content for relevance to the topic. Think of a handbook not as a ready-to-eat meal, but as a cookbook of ingredients that enables you to create the most fitting experience for the occasion.

RESOURCE Materials designed to support group learning. They come in many forms: a complete, ready-to-use exercise (such as a game); a comprehensive resource on one topic (such as conflict management) containing a variety of methods and approaches; or a collection of like-minded activities (such as icebreakers) on multiple subjects and situations.

TRAINING PACKAGE An entire, ready-to-use learning program that focuses on a particular topic or skill. All packages comprise a guide for the facilitator/trainer and a workbook for the participants. Some packages are supported with additional media—such as video—or learning aids, instruments, or other devices to help participants understand concepts or practice and develop skills.

- *Facilitator/trainer's guide* Contains an introduction to the program, advice on how to organize and facilitate the learning event, and step-by-step instructor notes. The guide also contains copies of presentation materials—handouts, presentations, and overhead designs, for example—used in the program.

- *Participant's workbook* Contains exercises and reading materials that support the learning goal and serves as a valuable reference and support guide for participants in the weeks and months that follow the learning event. Typically, each participant will require his or her own workbook.

ELECTRONIC CD-ROMs and web-based products transform static Pfeiffer content into dynamic, interactive experiences. Designed to take advantage of the searchability, automation, and ease-of-use that technology provides, our e-products bring convenience and immediate accessibility to your workspace.

METHODOLOGIES

CASE STUDY A presentation, in narrative form, of an actual event that has occurred inside an organization. Case studies are not prescriptive, nor are they used to prove a point; they are designed to develop critical analysis and decision-making skills. A case study has a specific time frame, specifies a sequence of events, is narrative in structure, and contains a plot structure—an issue (what should be/have been done?). Use case studies when the goal is to enable participants to apply previously learned theories to the circumstances in the case, decide what is pertinent, identify the real issues, decide what should have been done, and develop a plan of action.

ENERGIZER A short activity that develops readiness for the next session or learning event. Energizers are most commonly used after a break or lunch to stimulate or refocus the group. Many involve some form of physical activity, so they are a useful way to counter post-lunch lethargy. Other uses include transitioning from one topic to another, where "mental" distancing is important.

EXPERIENTIAL LEARNING ACTIVITY (ELA) A facilitator-led intervention that moves participants through the learning cycle from experience to application (also known as a Structured Experience). ELAs are carefully thought-out designs in which there is a definite learning purpose and intended outcome. Each step—everything that participants do during the activity—facilitates the accomplishment of the stated goal. Each ELA includes complete instructions for facilitating the intervention and a clear statement of goals, suggested group size and timing, materials required, an explanation of the process, and, where appropriate, possible variations to the activity. (For more detail on Experiential Learning Activities, see the Introduction to the *Reference Guide to Handbooks and Annuals*, 1999 edition, Pfeiffer, San Francisco.)

GAME A group activity that has the purpose of fostering team spirit and togetherness in addition to the achievement of a pre-stated goal. Usually contrived—undertaking a desert expedition, for example—this type of learning method offers an engaging means for participants to demonstrate and practice business and interpersonal skills. Games are effective for team building and personal development mainly because the goal is subordinate to the process—the means through which participants reach decisions, collaborate, communicate, and generate trust and understanding. Games often engage teams in "friendly" competition.

ICEBREAKER A (usually) short activity designed to help participants overcome initial anxiety in a training session and/or to acquaint the participants with one another. An icebreaker can be a fun activity or can be tied to specific topics or training goals. While a useful tool in itself, the icebreaker comes into its own in situations where tension or resistance exists within a group.

INSTRUMENT A device used to assess, appraise, evaluate, describe, classify, and summarize various aspects of human behavior. The term used to describe an instrument depends primarily on its format and purpose. These terms include survey, questionnaire, inventory, diagnostic, survey, and poll. Some uses of instruments include providing instrumental feedback to group members, studying here-and-now processes or functioning within a group, manipulating group composition, and evaluating outcomes of training and other interventions.

Instruments are popular in the training and HR field because, in general, more growth can occur if an individual is provided with a method for focusing specifically on his or her own behavior. Instruments also are used to obtain information that will serve as a basis for change and to assist in workforce planning efforts.

Paper-and-pencil tests still dominate the instrument landscape with a typical package comprising a facilitator's guide, which offers advice on administering the instrument and interpreting the collected data, and an initial set of instruments. Additional instruments are available separately. Pfeiffer, though, is investing heavily in e-instruments. Electronic instrumentation provides effortless distribution and, for larger groups particularly, offers advantages over paper-and-pencil tests in the time it takes to analyze data and provide feedback.

LECTURETTE A short talk that provides an explanation of a principle, model, or process that is pertinent to the participants' current learning needs. A lecturette is intended to establish a common language bond between the trainer and the participants by providing a mutual frame of reference. Use a lecturette as an introduction to a group activity or event, as an interjection during an event, or as a handout.

MODEL A graphic depiction of a system or process and the relationship among its elements. Models provide a frame of reference and something more tangible, and more easily remembered, than a verbal explanation. They also give participants something to "go on," enabling them to track their own progress as they experience the dynamics, processes, and relationships being depicted in the model.

ROLE PLAY A technique in which people assume a role in a situation/scenario: a customer service rep in an angry-customer exchange, for example. The way in which the role is approached is then discussed and feedback is offered. The role play is often repeated using a different approach and/or incorporating changes made based on feedback received. In other words, role playing is a spontaneous interaction involving realistic behavior under artificial (and safe) conditions.

SIMULATION A methodology for understanding the interrelationships among components of a system or process. Simulations differ from games in that they test or use a model that depicts or mirrors some aspect of reality in form, if not necessarily in content. Learning occurs by studying the effects of change on one or more factors of the model. Simulations are commonly used to test hypotheses about what happens in a system—often referred to as "what if?" analysis—or to examine best-case/worst-case scenarios.

THEORY A presentation of an idea from a conjectural perspective. Theories are useful because they encourage us to examine behavior and phenomena through a different lens.

TOPICS

The twin goals of providing effective and practical solutions for workforce training and organization development and meeting the educational needs of training and human resource professionals shape Pfeiffer's publishing program. Core topics include the following:

Leadership & Management

Communication & Presentation

Coaching & Mentoring

Training & Development

E-Learning

Teams & Collaboration

OD & Strategic Planning

Human Resources

Consulting

What will you find on pfeiffer.com?

- The best in workplace performance solutions for training and HR professionals

- Downloadable training tools, exercises, and content

- Web-exclusive offers

- Training tips, articles, and news

- Seamless on-line ordering

- Author guidelines, information on becoming a Pfeiffer Affiliate, and much more

Discover more at www.pfeiffer.com